COMMUNICATIONS ENGINEERING

COMMUNICATIONS ENGINEERING
Essentials for Computer Scientists and Electrical Engineers

R. C. T. Lee

Department of Computer Science
National Chi Nan University, Taiwan

Mao-Ching Chiu

Department of Electrical Engineering
National Chung Cheng University, Taiwan

Jung-Shan Lin

Department of Electrical Engineering
National Chi Nan University, Taiwan

IEEE PRESS
IEEE Communications Society, Sponsor

John Wiley & Sons (Singapore) Pte Ltd

Other Wiley Editorial Offices

John Wiley & Sons Inc., 111 River Street, Hoboken, NJ 07030, USA

Jossey-Bass, 989 Market Street, San Francisco, CA 94103-1741, USA

Wiley-VCH Verlag GmbH, Boschstr. 12, D-69469 Weinheim, Germany

John Wiley & Sons Australia Ltd, 42 McDougall Street, Milton, Queensland 4064, Australia

John Wiley & Sons Ltd, The Atrium, Southern Gate, Chichester, West Sussex PO19 8SQ, UK

John Wiley & Sons Canada Ltd, 6045 Freemont Blvd, Mississauga, ONT, L5R 4J3, Canada

Wiley also publishes its books in a variety of electronic formats. Some content that appears in print may not be
available in electronic books.

IEEE Communications Society, Sponsor
COMMS-S Liaison to IEEE Press, Mostafa Hashem Sherif

Anniversary Logo Design: Richard J. Pacifico

Library of Congress Cataloging-in-Publication Data

Communications engineering : essentials for computer scientists and
electrical engineers / R. C. T. Lee, Mao-Ching Chiu and Jung-Shan Lin.
   p. cm.
  Includes bibliographical references and index.
  ISBN 978-0-470-82245-6 (cloth)
1. Telecommunication.   I. Lee, R. C. T.
  TK5101. C65825 2007
  621.382– dc22   2007018943

ISBN 978-0-470-82245-6 (HB)

Typeset in 10/12 pt Times by Thomson Digital, India
Printed and bound in Singapore by Markono Print Media Pte Ltd.
This book is printed on acid-free paper responsibly manufactured from sustainable forestry
in which at least two trees are planted for each one used for paper production.

Contents

Preface

It may be said that our world is witnessing an amazing development of communication technology. Billions of people can watch the World Soccer Game at the same time. We can watch more than 60 channels of TV broadcast in our homes. Our computers are almost all connected to the Internet and we can easily exchange information through it. Besides, we can listen to beautiful music broadcasts via our computers and our speakers may possibly be linked to the tuners wirelessly.

Traditionally, communication engineering is a field inside electrical engineering. It is still true that most of the critical technologies of communication are related to electrical engineering. Although we are dealing with digital data most of the time, we still have to know frequency and bandwidth which are entirely within the realm of analog signal processing. Besides, when we transmit digital signals, such as in the case of mobile phone, we still need the help of radio frequency technologies.

But, as communication technologies become more and more advanced, a communication system cannot be realized without the help of computer scientists. Not mentioning the large systems used in those telephone companies, even for producing a mobile phone, a large amount of software systems must be developed. Suddenly, there is a big problem, most of programmers do not know what the electrical engineers are talking about. When they talk about terms such as QPSK, RF and OFDM, the poor computer scientists just sit there and do nothing.

On the other hand, computer scientists will gradually feel the need to understand how communication works. In the old days, a computer was a computer. The facilities making a computer connecting to the outside world was not an integral part of the computer. These days, since most computers are connected to the Internet, communication capability becomes a part of the computer, not something added on. If a computer scientist is ignorant of communication technology, it will be hard for him to design computer.

But, we must admit that most computer science students do not understand communication technology very well. For instance, they do not know how bits can be mixed and sent out, why we still talk about bandwidth when we only send pulses and no carrier frequency is involved and why we need the so-called RF technology in wireless communication.

This book was written for both electrical engineering and computer science students. Communication is a very complicated engineering technology. Since this is introductory textbook, we only discuss the key concepts of communication technologies, namely those critical and basic ones.

These days, we seem to be digitizing all kinds of things. We have digital TV, digital radio and almost all of the data we send are digital data. Most of students think that analog signals are ancient and out-of-date while digital signals are modern and advanced. Yet, when we send these digital data wirelessly, we always send them with the help of radio frequency signals which are analog. Thus, we talk about carrier frequency and bandwidth. Even when we send them through wires in the form of pulses where no carrier frequency is involved, we still talk about bandwidth. After all, we know that in order to have high bit rate, we must have broadband system. Why?

This book, in some sense, brings the attention of students back to analog signals. We make clear to the students that digital signals are after all also analog signals. A pulse consists of a set of cosine functions. The narrower the pulse is, the more cosine functions it contains and thus the larger the bandwidth. Thus, even when we send pulses, we have to be concerned with bandwidth.

We have tried our best to give the reader physical meaning of many terms as much as possible. A typical example is the discussion of Fourier transform. Most students are totally confused about the existence of negative frequencies in Fourier transform. In this book, this is explained clearly. Analog modulation is another example. We not only illustrate the mechanism of analog modulation, we also explain clearly why analog modulation is necessary.

We realize that communication technologies are quite involved and by no means easy to understand. Therefore we took pains to explain every concept in detail. The Fourier transform is introduced in almost every textbook in communication. That amplitude modulation will lift the frequencies of a signal is explained by using four different methods; some of them rather easy to understand. We are proud that we have presented many figures and tables. There are 142 figures and 14 tables in this book.

Only basic trigonometry and calculus are required for reading this book. Some parts of this book may be found to be too difficult to some students. Then the instructor can ignore them.

To help the students have some feeling about existing communications systems, we tried to introduce briefly many such systems. For instance, AM and FM broadcasting, Ham radio, walkie-talkie, mobile phone, hand-held phone, T1 to T4, ADSL, digital radio, Blue-tooth and so on. In each system, we talked about the modulation scheme, carrier frequency and bandwidth which are important parameters.

It is our experience that this book can be used in a one-semester course. Since no sophisticated mathematics are required, this book can be taught to college students freshmen above.

Solutions manual, slides, lecture notes and an instructors manual can be found at: http://www.wiley.com/go/rctlee

1

An Overview of Computer Communications

Among all the modern technologies available to us, communications technology is perhaps one of the most puzzling. We use radio and television sets all the time but perhaps struggle to understand how these things work. We might also be quite puzzled when we use our mobile phones. When we make a call, messages are sent out to, say, a base station – which we vaguely know. Yet there must be many other people close by who are using the mobile phone system, so that base station must be receiving a mixture of signals. How can it recover all the *individual* signals? A lot of us use ADSL these days to connect a computer at home to the Internet. Again, it might be hard to understand what ADSL is and how it works.

Actually, communications technologies are not that hard. The first thing we have to understand is the nature of signals. We usually see a signal from the viewpoint of time, but it is important to know that a signal actually is composed of a set of *cosine functions*. How do we know? We know this through the use of the *Fourier transform*. This transforms a signal in the *time domain* into the *frequency domain*. It is an exceedingly powerful tool. In fact, we could say that modern telecommunications would not be possible without the Fourier transform.

With the Fourier transform we can view any signal from the viewpoint of its contituent frequencies. When the ordinary person talks about digital signals, he or she might not relate the pulses with frequencies. The person can easily understand the *data rate* of digital signals, but we will show that digital pulses are related to frequencies.

Let us consider radio broadcasting, which we are all familiar with. There are many radio transmitting stations. They all broadcast either the human voice or music, or both. We must give each radio station a unique frequency to broadcast. That frequency is called the *carrier frequency*. Compared with the frequencies associated with voices or music, these carrier frequencies are much higher. The radio stations, through some mechanism, attach the voice or music signals to the higher carrier frequencies. This mechanism is called *modulation*.

The air is thus filled with all kinds of signal corresponding to different frequencies. Our radio receives some of these. Actually we *tune* our radio so that at a particular time it picks out one particular carrier frequency. In this way we can hear the broadcast clearly. This process in the radio is the opposite of modulation and is called *demodulation*.

Through use of the Fourier transform, it can be shown that the human voice and music consist of relatively *low frequencies*. Can low-frequency signals be broadcast directly? No,

Communications Engineering. R.C.T. Lee, Mao-Ching Chiu and Jung-Shan Lin
© 2007 John Wiley & Sons (Asia) Pte Ltd

because low-frequency signals have *large wavelength*. Unfortunately, the length of an antenna is proportional to the wavelengths of the signals it can handle. Thus, to broadcast low-frequency signals we would need a very large antenna, which is not practical. After introducing the much higher carrier frequency, we can then use reasonably sized antennas to broadcast and to receive.

In the above, our modulation technique is assumed to be applied to *analogue signals*, so this is called *analogue modulation*. The carrier frequencies are often called *radio frequencies* (RF for short). Actually, for reasons that need not bother us here, 'RF' today often refers to high frequencies that are no longer restricted to radio broadcasting.

What about *digital signals*? Data are almost always represented by digital signals. Digital signals are often thought of as a *sequence of pulses*. We must note that, in the wireless environment, we cannot transmit pulses except in very unusual cases. This is because, in the wireless environment, only *electromagnetic waves* are transmitted. What we can do is to transmit a *sinusoidal signal* within that short period of duration of a single pulse. This sinusoidal signal is a carrier signal and its frequency is called the carrier frequency. In this way, each user can again be uniquely identified by its carrier frequency. This kind of modulation is called *digital modulation*.

Digital modulation is more interesting than that. We can transmit more than one *bit* at the same time. This will make the communication more efficient because two bits can be sent. Note that the first bit may be 1 or 0 and the second may also be 1 or 0. Thus there are four possible cases for the receiver to determine at any instant. How can the receiver do the job of distinguishing the value of the bit i ($i = 1, 2$)? This is a very fundamental problem which we must be able to deal with.

Fortunately, there is a ready-made answer. Mathematics tells us that as long as the signals are *orthogonal* to each other, they can be mixed and later recovered easily by using the *inner-product operation*. The inner-product operation is thus very important, so in this book it is introduced quite early.

If two bits can be mixed together, a larger number of bits can also be mixed. Thus the *orthogonal frequency-division method* (OFDM) is capable of mixing 256 bits together. In this method, each bit is represented by a cosine function with a distinct frequency. It can be easily shown that these functions are orthogonal and thus can be recovered. For OFDM systems, we shall use the *discrete Fourier transform* in this book, so that the transmission can be done efficiently. Our ADSL system is based on this ODFM method.

In communications, it is natural to have the situation in which several users want to send signals to the same receiver simultaneously. A typical case is the mobile phone system whereby many callers around one base station are using the same base station. How can the receiver distinguish between the users? One straightforward method to distinguish data generated by different users is to use the *frequency-division multiple-access* (FDMA) method. We can also use the *time-division multiple-access* (TDMA) method, in which data are divided according to time slots. Another very interesting technique is the *code-division multiple-access* (CDMA) method. In the CDMA method, different users use different codes to represent the values of bits. Each code can be considered as a function, and they can be recovered because the functions corresponding to different codes are orthogonal.

A very important concept in communications is *bandwidth*. Consider the human voice case. Experimental results show that the human voice contains frequencies mainly from 0 to

5000 hertz (Hz), which means that the bandwidth of any system transmitting this voice signal must be larger than 5000 Hz. Let us denote the carrier frequency by f_c. After analogue modulation is done, the frequencies now range from $(f_c - 5)$ kHz to $(f_c + 5)$ kHz. We will then say that the bandwidth of the signal is 10 kHz. Unfortunately, a large bandwidth of a communication system occupies more resources and requires more sophisticated electronic circuitry. Thus we shall often be mentioning the concept of bandwidth.

Now consider a cable TV system. The cable is used to transmit a number of TV signals. Assume that there are N television stations and that the bandwidth of each TV signal is W. The bandwidth of the TV cable must therefore be larger than NW. For large N and W, the bandwidth must be quite large, and this is why we often call this kind of system a *broadband system*.

It is easy to understand the bandwidth of analogue signals, but digital signals also have a bandwidth associated with them. Every digital signal can be seen as a sequence of pulses. Each pulse has a *pulse width*. When we introduce the Fourier transform, we will show that a short pulse width will occupy a wide band of frequencies. Further, a high data rate will necessarily mean a short pulse width, and consequently a large bandwidth. Thus, we may say that if we want to transmit a large number of bits in a short time, we must have a communications system with a large bandwidth. So, even when we send pure digital data, the concept of bandwidth is still important.

We do not, of course, want a transmission mechanism that requires a very large bandwidth. That would be very costly. On the other hand, a very narrow bandwidth transmission has the disadvantage that it is easy for intruders to penetrate. For security reasons, sometimes we would like to widen the bandwidth. We will introduce the concept of *spread spectrum technology*, by which the bandwidth of a system is widened which will make it more secure.

Spread spectrum technology is designed not only to make a system more secure. *Frequency hopping*, for instance, is a spread spectrum technology that allows a transceiver (transmitter/receiver) to communicate with many other transceivers. Imagine, for example, that we have equipment in a laboratory which are connected to many devices. Each device sends data to the equipment from time to time, and the equipment will have to send instructions to these devices very often. Frequency hopping allows this to happen.

Finally, *coding* is something that we must understand. There are two kinds. Removing redundant data is called *source coding*, and adding redundant data to correct errors is called *channel coding*. Both are introduced in this book.

Further Reading

- For a detailed discussion of ADSL, see [B00].
- For the history of communications, see [L95].
- For a detailed discussion of communication networks, see [T95].

2

Signal Space Representation

This chapter discusses a very important topic. Suppose we have to mix signals together, how can we then extract them correctly? Since these signals are basically sinusoidal, it is important for the reader to consult the formulas in Appendix A (which contains most of the formulas frequently used in the theory of communications) whenever he or she is not familiar with the equations used in the derivations.

An important concept in communications will now be introduced: Let us consider the case where we have the following periodic function (the term 'periodic function' will be defined fully in the next chapter.):

$$f(t) = a\cos(2\pi f_c t) + b\sin(2\pi f_c t).$$

The period of this signal is $T = 1/f_c$. Suppose a receiver receives the above signal and, for some reason, it wants to determine the values of a and b. What can we do? To determine a, we may perform the following calculation:

$$\int_0^T f(t)\cos(2\pi f_c t)dt$$

$$= a\int_0^T \cos^2(2\pi f_c t)dt + b\int_0^T \sin(2\pi f_c t)\cos(2\pi f_c t)dt$$

$$= \frac{a}{2}\int_0^T (1 + \cos(4\pi f_c t))dt + \frac{b}{2}\int_0^T \sin(4\pi f_c t)dt$$

$$= \frac{aT}{2} + \frac{a}{2(4\pi f_c)}\sin(4\pi f_c t)\Big|_0^T - \frac{b}{2(4\pi f_c)}\cos(4\pi f_c t)\Big|_0^T$$

$$= \frac{aT}{2} + \frac{a}{8\pi f_c}(0 - 0) - \frac{b}{8\pi f_c}(1 - 1)$$

$$= \frac{aT}{2}.$$

Communications Engineering. R.C.T. Lee, Mao-Ching Chiu and Jung-Shan Lin
© 2007 John Wiley & Sons (Asia) Pte Ltd

Thus:

$$a = \frac{2}{T} \int_0^T f(t) \cos(2\pi f_c t) dt.$$

Similarly, the following expression gives us the value of b:

$$\int_0^T f(t) \sin(2\pi f_c t) dt.$$

That the values of a and b can be determined in the above manner is due to the *orthogonality of cosine and sine functions*. This concept will be used throughout the book, so we devote this entire chapter to it. First, we shall discuss the vector space concept.

2.1 The Vector Space

In an n-dimensional space, a vector $\mathbf{v}$ is an n-tuple $(v_1, v_2, \cdots, v_n)$. Let us first define the inner product of two vectors. Let $\mathbf{v}_1 = (v_{11}, v_{12}, \cdots, v_{1n})$ and $\mathbf{v}_2 = (v_{21}, v_{22}, \cdots, v_{2n})$. The *inner product* of these two vectors is

$$\mathbf{v}_1 \cdot \mathbf{v}_2 = \sum_{i=1}^{n} v_{1i} v_{2i}. \tag{2-1}$$

The norm of a vector $\mathbf{v}$, denoted as $||\mathbf{v}||$, is defined as $(\mathbf{v} \cdot \mathbf{v})^{1/2}$. Two vectors $\mathbf{v}_1$ and $\mathbf{v}_2$ are said to be *orthogonal* if the inner product between them is 0.

Example 2-1
Consider $\mathbf{v}_1 = (1, 1)$ and $\mathbf{v}_2 = (1, -1)$. The inner product of these vectors, $\mathbf{v}_1 \cdot \mathbf{v}_2 = (1 \times 1) + (1 \times (-1)) = 1 - 1 = 0$. Therefore these vectors are orthogonal to each other.

A set of vectors are orthonormal if each pair of these vectors are orthogonal to each other and the norm of each vector is 1. For instance, $(1,0)$ and $(0,1)$ are orthonormal and $(1/\sqrt{2}, 1/\sqrt{2})$ and $(1/\sqrt{2}, -1/\sqrt{2})$ are also orthonormal.

Let us consider two vectors $\mathbf{v}_1$ and $\mathbf{v}_2$ which are orthogonal and $\mathbf{v} = a\mathbf{v}_1 + b\mathbf{v}_2$. Then a can be found by $\mathbf{v} \cdot \mathbf{v}_1$ because

$$\mathbf{v} \cdot \mathbf{v}_1 = a(\mathbf{v}_1 \cdot \mathbf{v}_1) + b(\mathbf{v}_2 \cdot \mathbf{v}_1)$$
$$= a||\mathbf{v}_1||^2 + b \cdot 0$$
$$= a||\mathbf{v}_1||^2$$

Thus:

$$a = \frac{\mathbf{v} \cdot \mathbf{v}_1}{||\mathbf{v}_1||^2}.$$

The value of b can be found in a similar way.

Example 2-2

Consider vectors $\mathbf{v}_1 = (1/\sqrt{2}, 1/\sqrt{2})$ and $\mathbf{v}_2 = (1/\sqrt{2}, -1/\sqrt{2})$. Let $\mathbf{v} = (7/\sqrt{2}, -1/\sqrt{2})$. Suppose that $\mathbf{v} = a\mathbf{v}_1 + b\mathbf{v}_2$. Then a can be found as follows:

$$a = \frac{\mathbf{v} \cdot \mathbf{v}_1}{\|\mathbf{v}_1\|^2} = \frac{\left(\dfrac{7}{2} - \dfrac{1}{2}\right)}{1} = 3.$$

Similarly, b can be found by

$$b = \frac{\mathbf{v} \cdot \mathbf{v}_2}{\|\mathbf{v}_2\|^2} = \frac{\left(\dfrac{7}{2} + \dfrac{1}{2}\right)}{1} = 4.$$

In general, if we have $\mathbf{v} = \sum_{i=1}^{n} a_i \mathbf{v}_i$, where the $\mathbf{v}_i$ are orthogonal with one another, a_i can be found as follows:

$$a_i = \frac{\mathbf{v} \cdot \mathbf{v}_i}{\|\mathbf{v}_i\|^2}. \tag{2-2}$$

2.2 The Signal Space

Signals are functions of time. Let us denote a signal by $\phi(t)$. Given two signals $\phi_i(t)$ and $\phi_j(t)$, the inner product of these signals, denoted as $\langle \phi_i(t), \phi_j(t) \rangle$ is defined as follows:

$$\langle \phi_i(t), \phi_j(t) \rangle = \int_a^b \phi_i(t)\phi_j(t)dt. \tag{2-3}$$

Note that the inner product of two signals, defined as in (2-3), is also a function of the two parameters a and b. Different values of a and b will give different results. To emphasize this point, in this book, whenever it is needed, we shall sometimes say that this is an inner product of $\phi_1(t)$ and $\phi_2(t)$ with respect to interval $\{a, b\}$.

Two signals are said to be *orthogonal* if their inner product is 0. The norm of a signal $\phi(t)$, denoted as $\|\phi(t)\|$, is defined as $(\langle \phi(t), \phi(t) \rangle)^{1/2}$. A set of signals are called *orthonormal* if they are orthogonal to one another and their norms are all equal to 1.

In the rest of this section, whenever parameters k and n are used as coefficients, they denote positive integers, unless otherwise stated. Let us now give several examples.

Example 2-3

Let $\phi_1(t) = \sin(2\pi k f_c t)$ and $\phi_2(t) = \sin(2\pi n f_c t)$. The inner product of these signals will depend on the interval of integration and whether $k = n$ or $k \neq n$. In the following, we assume that the interval of integration is $\{0, T\}$ where $T = 1/f_c$.

Case 1: $k \neq n$. In this case, we use Equation A-10:

$$
\begin{aligned}
\langle \phi_1(t), \phi_2(t) \rangle &= \int_0^T \sin(2\pi k f_c t) \sin(2\pi n f_c t) dt \\
&= \int_0^T \frac{1}{2} [\cos(2\pi(k-n)f_c t) - \cos(2\pi(k+n)f_c t)] dt \\
&= \frac{1}{2} \left[\frac{\sin(2\pi(k-n)f_c t)}{(2\pi(k-n)f_c)} - \frac{\sin(2\pi(k+n)f_c t)}{(2\pi(k+n)f_c)} \right] \Bigg|_{t=0}^{t=T} \\
&= 0
\end{aligned}
\tag{2-4}
$$

Case 2: $k = n \neq 0$. In this case, we use Equation A-17:

$$
\langle \phi_1(t), \phi_2(t) \rangle = \frac{1}{2} \int_0^T (1 - \cos(2\pi(2k)f_c t)) dt = \frac{1}{2} \Big|_{t=0}^{t=T} = \frac{T}{2}.
\tag{2-5}
$$

In summary, we have:

$$
\langle \sin(2\pi k f_c t), \sin(2\pi n f_c t) \rangle = 0 \quad \text{if} \quad k \neq n,
\tag{2-6}
$$

$$
\langle \sin(2\pi k f_c t), \sin(2\pi n f_c t) \rangle = \frac{T}{2} \quad \text{if} \quad k = n \neq 0.
\tag{2-7}
$$

Identical results hold for $\phi_1(t) = \cos(2\pi k f_c t)$ and $\phi_2(t) = \cos(2\pi n f_c t)$.

Example 2-4

Let $\phi_1(t) = \sin(2\pi k f_c t)$ and $\phi_2(t) = \cos(2\pi n f_c t)$.

Case 1: $k \neq n$. We use Equation A-11:

$$
\begin{aligned}
\langle \phi_1(t), \phi_2(t) \rangle &= \int_0^T \sin(2\pi k f_c t) \cos(2\pi n f_c t) dt \\
&= \int_0^T \frac{1}{2} (\sin(2\pi(k-n)f_c t) + \sin(2\pi(k+n)f_c t)) dt \\
&= -\frac{1}{2} \left[\frac{\cos(2\pi(k-n)f_c t)}{(2\pi(k-n)f_c)} + \frac{\cos(2\pi(k+n)f_c t)}{(2\pi(k+n)f_c)} \right]_{t=0}^{t=T} \\
&= 0.
\end{aligned}
\tag{2-8}
$$

Case 2: $k = n \neq 0$. We use Equation A-13:

$$
\langle \phi_1(t), \phi_2(t) \rangle = \frac{1}{2} \int_0^T (\sin(2\pi(2k)f_c t)) dt = 0.
\tag{2-9}
$$

Thus we have:

$$\langle \sin k(2\pi f_c t), \cos n(2\pi f_c t) \rangle = 0 \quad \text{for all } k \text{ and } n. \tag{2-10}$$

We may summarize the results from Examples 2-3 and 2-4 as follows:

$$\langle \sin(2\pi k f_c t), \sin(2\pi n f_c t) \rangle = 0 \text{ if } k \neq n \tag{2-11}$$

$$\langle \sin(2\pi k f_c t), \sin(2\pi n f_c t) \rangle = \frac{T}{2} \text{ if } k = n \neq 0 \tag{2-12}$$

$$\langle \cos(2\pi k f_c t), \cos(2\pi n f_c t) \rangle = 0 \text{ if } k \neq n \tag{2-13}$$

$$\langle \cos(2\pi k f_c t), \cos(2\pi n f_c t) \rangle = \frac{T}{2} \text{ if } k = n \neq 0 \tag{2-14}$$

$$\langle \sin(2\pi k f_c t), \cos(2\pi n f_c t) \rangle = 0 \text{ for all } k \text{ and } n. \tag{2-15}$$

Table 2-1 summarizes the above equations. As shown, sine functions, or cosine functions, are not orthogonal only to themselves.

Some examples are summarized in Figure 2-1, which illustrates four cases: $\sin(x)\cos(3x)$, $\sin(x)\sin(3x)$, $\sin(x)\cos(x)$ and $\sin(x)\sin(x)$. The reader should examine each case carefully. For the first three cases, the total area above the x-axis is equal to the total area below the x-axis. Consequently, integration of the functions over the period from 0 to 2π yields 0 for these three cases. For the last case, there is no area below the x-axis. Thus, the integration yields a non-zero result. These results are compatible with Equations (2-11) to (2-15).

The above results can be explained from another point of view. Note that the integration of a sinusoidal function over its period is 0. An examination of Equations (A-9) to (A-13) in Appendix A shows that $\cos\alpha\cos\beta$, $\sin\alpha\sin\beta$, $\sin\alpha\cos\beta$ and $\cos\alpha\sin\beta$ are all equivalent to summations of cosine and sine functions. If we perform an integration on them over the period, the result will be zero. On the other hand, Equations (A-17) and (A-18) show that there are constants in the formulas for $\sin^2\alpha$ and $\cos^2\alpha$. Thus integration over them will produce a non-zero result.

Suppose that our signal is represented by the following formula:

$$T = 1/S_c, f(t) = \sum_{k=1}^{n} a_k \cos(2\pi k f_c t) + \sum_{k=1}^{n} b_k \sin(2\pi k f_c t) \tag{2-16}$$

Table 2-1   The orthogonality of sinusoidal functions

Inner product	$\cos(2\pi k f_0 t)$	$\cos(2\pi n f_0 t)$	$\sin(2\pi k f_0 t)$	$\sin(2\pi n f_0 t)$
$\cos(2\pi k f_0 t)$	$\frac{T}{2}$	0	0	0
$\cos(2\pi n f_0 t)$	0	$\frac{T}{2}$	0	0
$\sin(2\pi k f_0 t)$	0	0	$\frac{T}{2}$	0
$\sin(2\pi n f_0 t)$	0	0	0	$\frac{T}{2}$

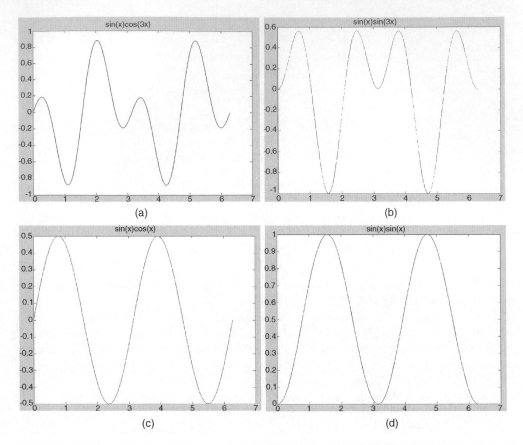

Figure 2-1 Illustration of four functions: (a) $\sin(x)\cos(3x)$, (b)$\sin(x)\sin(3x)$, (c)$\sin(x)\cos(x)$, and (d)$\sin(x)\sin(x)$

and we are asked to determine a_k and b_k. To do this, we may use the above equations concerned with inner products of cosine and sine functions. That is, to determine a_k, we perform the following:

$$\langle f(t), \cos(2\pi k f_c t) \rangle = \int_0^T f(t)\cos(2\pi k f_c t)dt. \tag{2-17}$$

From Equations (2-11) to (2-15), we can easily see that all terms in Equation (2-17) vanish except a_k. Thus, from Equation (2-14), we have:

$$a_k = \frac{2}{T}\langle f(t), \cos(2\pi k f_c t) \rangle = \frac{2}{T}\int_0^T f(t)\cos(2\pi k f_c t)dt. \tag{2-18}$$

Similarly, b_k can be found by the following formula:

$$b_k = \frac{2}{T} \langle f(t), \sin(2\pi k f_c t) \rangle = \frac{2}{T} \int_0^T f(t) \sin(2\pi k f_c t) dt. \tag{2-19}$$

The above result will be very useful when we introduce the concepts of Fourier series, Fourier transform, amplitude modulation, double sideband modulation, single sideband modulation, QPSK and OFDM etc. in later chapters.

Example 2-5

Consider the case where $\phi_1(t) = \sqrt{\frac{2}{T}}\cos(2\pi f_1 t)$ and $\phi_2(t) = \sqrt{\frac{2}{T}}\cos(2\pi f_2 t)$ under the condition that $0 \le t < T$ and

$$f_i = \frac{n_c + i}{T} \quad \text{for some integer } n_c \text{ and } i = 1, 2. \tag{2-20}$$

$$\langle \phi_1(t), \phi_2(t) \rangle = \frac{2}{T} \int_0^T \cos(2\pi f_1 t)\cos(2\pi f_2 t) dt$$

$$= \frac{1}{T} \int_0^T (\cos(2\pi(f_1 + f_2)t) + \cos(2\pi(f_1 - f_2)t)) dt. \tag{2-21}$$

Substituting Equation (2-20) into Equation (2-21), we have:

$$\langle \phi_1(t), \phi_2(t) \rangle = \frac{1}{T} \left(\int_0^T \cos\left(2\pi\left(\frac{2n_c + 3}{T}t\right)\right) dt + \int_0^T \cos\left(2\pi\left(\frac{-1}{T}\right)t\right) dt \right). \tag{2-22}$$

In the above equation, for both the first and the second integrations, an integer number of periods are involved. Therefore both terms are zero. We thus have:

$$\langle \phi_1(t), \phi_2(t) \rangle = \langle \cos(2\pi f_1 t), \cos(2\pi f_2 t) \rangle = 0 \tag{2-23}$$

under the condition that Equation (2-20) is satisfied.

We can also prove that, for $i = 1,2$:

$$\langle \phi_i(t), \phi_i(t) \rangle = 1. \tag{2-24}$$

There is another way of looking at the problem. Let $f_c = 1/T$. Then $f_1 = (n_c + 1)f_c$ and $f_2 = (n_c + 2)f_c$. We may say that $f_1 = kf_c$, $f_2 = nf_c$ and $k \ne n$. Thus, according to Equation (2-13), we can easily derive Equation (2-23).

Suppose that a signal is represented as $f(t) = a\phi_1(t) + b\phi_2(t)$, where $\langle \phi_1(t), \phi_2(t) \rangle = 0$ and $\langle \phi_i(t), \phi_i(t) \rangle = 1$ for $i = 1, 2$. Then, it can be easily seen that a and b can be found through the following formulas:

$$a = \int_0^T f(t)\phi_1(t)dt$$

$$b = \int_0^T f(t)\phi_2(t)dt. \tag{2-25}$$

The above result will be useful when we introduce binary frequency-shift keying in Chapter 5.

Example 2-6
Let us consider the two signals illustrated in Figure 2-2(a) and 2-2(b). The multiplication of these two signals is shown in Figure 2-2(c), and the integration of the signal in 2-2(c) is shown in 2-2(d). One can see that these two signals are orthogonal because $\int_0^T \phi_1(t)\phi_2(t)dt = 0$.

Example 2-7
Consider two signals similar to those in Example 2-6. As shown in Figure 2-3, these two signals are again orthogonal.

In the above examples, we can also consider these two signals as vectors because they assume discrete values. In Example 2-6, the two signals can be represented as follows:

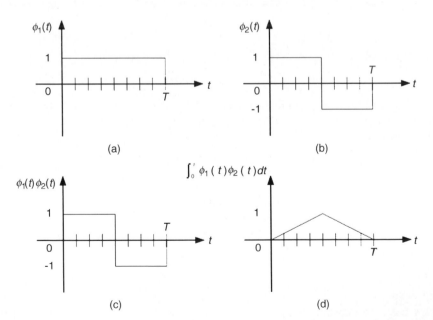

Figure 2-2   Two orthogonal signals: (a) $\phi_1(t)$, (b) $\phi_2(t)$, (c) $\phi_1(t)\phi_2(t)$, and (d) $\int_0^t \phi_1(t)\phi_2(t)dt$

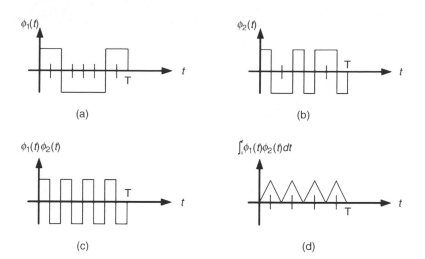

Figure 2-3 Another two signals: (a) $\phi_1(t)$, (b) $\phi_2(t)$, (c) $\phi_1(t)\phi_2(t)$, and (d) $\int_0^t \phi_1(t)\phi_2(t)dt$

$\mathbf{v}_1 = (1,1)$ and $\mathbf{v}_2 = (1,-1)$. The inner product of these vectors is

$$\mathbf{v}_1 \cdot \mathbf{v}_2 = (1 \times 1) + (1 \times (-1)) = 1 + (-1) = 0.$$

Similarly, the signals in Example 2-7 can be represented as $\mathbf{v}_1 = (1,1,-1,-1,-1,-1,1,1)$ and $\mathbf{v}_2 = (1,-1,-1,1,-1,1,1,-1)$. The inner product of these two vectors is

$$
\begin{aligned}
\mathbf{v}_1 \cdot \mathbf{v}_2 \\
&= (1 \times 1) + (1 \times (-1)) + ((-1) \times (-1)) + ((-1) \times 1) \\
&\quad + ((-1) \times (-1)) + ((-1) \times 1) + (1 \times 1) + (1 \times (-1)) \\
&= 1 + (-1) + 1 + (-1) + 1 + (-1) + 1 + (-1) \\
&= 0
\end{aligned}
$$

It is interesting to ask whether the signal space concept and the vector space concept are entirely different. Consider two signals denoted as $\phi_1(t)$ and $\phi_2(t)$. The inner product between $\phi_1(t)$ and $\phi_2(t)$ is defined to be $\int_a^b \phi_1(t)\phi_2(t)dt$. Suppose we do not have analytical formulas for $\phi_1(t)$ and $\phi_2(t)$. We can sample these two signals at time $t_1, t_2, \cdots, t_n$ from $t = a$ to $t = b$. Thus $\phi_1(t)$ is characterized as $\phi_1(t_1), \phi_1(t_2), \cdots, \phi_1(t_n)$ and $\phi_2(t)$ is characterized as $\phi_2(t_1), \phi_2(t_2), \cdots, \phi_2(t_n)$. Let $A = (\phi_1(t_1), \phi_1(t_2), \cdots, \phi_1(t_n))$ and $B = (\phi_2(t_1), \phi_2(t_2), \cdots, \phi_2(t_n))$. Then we can easily see that:

$$A \cdot B = \frac{b-a}{n} \sum_{i=1}^n \phi_1(t_i)\phi_2(t_i) = \int_a^b \phi_1(t)\phi_2(t)dt \quad \text{as} \quad n \to \infty.$$

Let us redraw Figure 2-3(c) as in 2-4(a). We can see that the integration of $\phi_1(t)\phi_2(t)$ is equivalent to finding the total area covered by the function $\phi_1(t)\phi_2(t)$. From $t = 0$ to $t = 1$, the area is $+1$. From $t = 1$ to $t = 2$, the area is -1. Thus the total area from $t = 0$ to $t = 2$ is

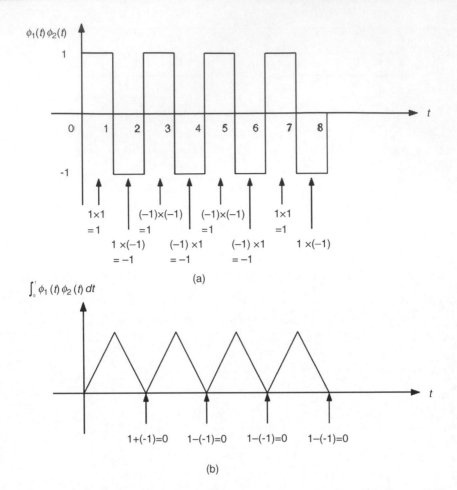

(a)

(b)

Figure 2-4 Comparison of the signal space inner product and vector space inner product: (a) vector space inner product, (b) signal space inner product

$+1 + (-1) = 0$. This exactly corresponds to the first two steps of the calculation of $\mathbf{v}_1 \cdot \mathbf{v}_2$ as shown above. Note that the first step of $\mathbf{v}_1 \cdot \mathbf{v}_2$ is $1 \times 1 = 1$ and the second step is $1 \times (-1) = -1$. The sum of the results of the first two steps is therefore $1 + (-1) = 0$. This process continues as indicated in Figure 2-4(b). Note that at the end, the result is 0.

The above result is useful when we introduce the CDMA mechanism which will be presented later.

2.3 Summary

In general, as will be shown later, a signal is often represented as follows:

$$f(t) = \sum_{i=1}^{n} a_i \phi_i(t)$$

where the ϕ_i are orthogonal. Our job is often to find the a_i. Because of the orthogonal properties of the ϕ_i, a_i can be found as follows:

$$a_i = \frac{1}{||\phi_i(t)||^2} \int_0^T f(t)\phi_i(t)dt.$$

This mechanism is used again and again in many chapters of this book.

Further Reading

- For more details about the background of linear algebra, see [AR05].
- For more details about the relations between signal and systems, see [C98].

Exercises

2.1 (a) Plot $\cos(2\pi(2)t)$ and $\sin(2\pi(3)t)$.
  (b) Explain why $\int_0^1 \cos(2\pi(2)t)dt = 0$ and $\int_0^1 \sin(2\pi(3)t)dt = 0$.
2.2 (a) Plot $\cos^2(2\pi t)$ and $\sin^2(2\pi t)$.
  (b) Explain why $\int_0^1 \cos^2(2\pi t)dt \neq 0$ and $\int_0^1 \sin^2(2\pi t)dt \neq 0$.
2.3 Complete the following equations:
  $\cos(\alpha + \beta) =$
  $\cos(\alpha - \beta) =$
  $\sin(\alpha + \beta) =$
  $\sin(\alpha - \beta) =$
2.4 Complete the following equations:
  $\cos\alpha \cos\beta =$
  $\sin\alpha \sin\beta =$
  $\sin\alpha \cos\beta =$
  $\cos\alpha \sin\beta =$
2.5 Complete the following equations:
  $\sin(2\alpha) =$
  $\cos(2\alpha) =$
  $1 - 2\sin^2(\alpha) = 2\cos^2(\alpha) - 1 =$
  $2\sin(\alpha)\cos(\alpha) =$
2.6 Complete the following:
  $\sin^2\alpha =$
  $\cos^2\alpha =$
2.7 Let $\mathbf{v}_1 = \left(\cos(2\pi(0)), \cos\left(2\pi\left(\frac{1}{4}\right)\right), \cos\left(2\pi\left(\frac{2}{4}\right)\right), \cos\left(2\pi\left(\frac{3}{4}\right)\right)\right)$
  and $\mathbf{v}_2 = \left(\sin(2\pi(0)), \sin\left(2\pi\left(\frac{1}{4}\right)\right), \sin\left(2\pi\left(\frac{2}{4}\right)\right), \sin\left(2\pi\left(\frac{3}{4}\right)\right)\right)$.
  Compute $\mathbf{v}_1 \cdot \mathbf{v}_2$. Explain the result.
2.8 (a) Let $\mathbf{v}_1 = \left(\frac{1}{2}, \frac{\sqrt{3}}{2}\right)$ and $\mathbf{v}_2 = \left(-\frac{\sqrt{3}}{2}, \frac{1}{2}\right)$. Show that these two vectors are orthonormal.
  (b) Let $\mathbf{v} = a\mathbf{v}_1 + b\mathbf{v}_2 = \left(\frac{2+3\sqrt{3}}{2}, \frac{2\sqrt{3}-2}{2}\right)$. Find a and b.
2.9 Let $\phi_1(t) = \cos(2\pi t)$ and $\phi_2(t) = \cos\left(2\pi t - \frac{\pi}{4}\right)$. Are $\phi_1(t)$ and $\phi_2(t)$ orthogonal to each other? Why?

2.10 Let $\phi_1(t) = \cos(2\pi t)$ and $\phi_2(t) = \cos\left(2\pi t - \frac{\pi}{2}\right)$. Are $\phi_1(t)$ and $\phi_2(t)$ orthogonal to each other? Why?

2.11 Let $V_1 = (1, -1, 1, -1, -1, 1, -1, 1)$ and $V_2 = (1, -1, -1, 1, -1, 1, 1, -1)$.
   (a) Prove that these two vectors are orthogonal to each other.
   (b) Let $V = (0, 0, 2, -2, 0, 0, -2, 2) = aV_1 + bV_2$. Determine a and b.

2.12 Consider the following vectors:

   $V_1 = (1, 1, 1, 1, 1, 1, 1, 1)$
   $V_2 = (1, -1, 1, -1, 1, -1, 1, -1)$
   $V_3 = (1, 1, -1, -1, 1, 1, -1, -1)$
   $V_4 = (1, -1, -1, 1, 1, -1, -1, 1)$.

   (a) Prove that these four vectors are orthogonal to each other.
   (b) Let $V = (-2, -2, 4, 0, -2, -2, 4, 0) = aV_1 + bV_2 + cV_3 + dV_4$. Determine a, b, c and d.

2.13 Let $\phi_1(t) \cos(2\pi t)$ and $\phi_2(t) \cos(6\pi t)$. Are these signals orthogonal to each other? Explain.

3

Fourier Representations of Signals

In this chapter, the concept of representing a signal in both time-domain and frequency-domain spaces is introduced and applied. Usually, a signal is described as a function of time. However, there are some amazing advantages if a signal can be expressed in the frequency domain. To represent a signal in both time and frequency domains results from the introduction of a series of harmonically related sinusoids. The study of signals and systems using sinusoidal representations is termed *Fourier transform analysis* after Jean Baptiste Joseph Fourier (1768–1830), a great French mathematician, for his contributions to the theory of representing functions as weighted superpositions of sinusoids. Fourier methods including Fourier series and Fourier transform for both continuous-time and discrete-time cases have widespread applications in almost every branch of engineering and science.

Before giving formal concepts related to the Fourier transform, let us first consider Figure 3-1. Imagine that we want to transmit this signal, which is denoted as $f(t)$. A straightforward way to do so is to perform a sampling and then to transmit the sampled points. That is, we transmit $f(t_1), f(t_2), \cdots, f(t_n)$. It can be easily seen that n should be as large as possible. In order to make sure that $f(t)$ is received accurately, a large amount of data has to be transmitted. This puts a heavy burden to the communications system.

The signal in Figure 3-1 is not sinusoidal. Suppose we apply the discrete Fourier transform to the signal in Figure 3-1 after sampling. The result is presented in Figure 3-2.

From Figure 3-2, it can be seen that the signal consists of two sinusoidal functions. One is of frequency 7 and with a small amplitude. The other is of frequency 13 and with a large amplitude. We shall show later that the frequencies 51 and 57 should be ignored. In fact, the discrete Fourier transform indicates that the signal in Figure 3-1, denoted as $f(t)$, can be expressed by the following equation:

$$f(t) = \cos(2\pi(7)t) + 3\cos(2\pi(13)t). \tag{3-1}$$

In other words, instead of transmitting $f(t_1), f(t_2), \cdots, f(t_n)$, we only have to send the frequencies 7 and 13, and their corresponding amplitudes, of the two cosine functions. The receiving end, after receiving these parameters, can reconstruct $f(t)$ by using Equation (3-1).

Let us now consider Figure 3-3. The result of applying discrete the Fourier transform to the signal is shown in Figure 3-4. From this we can see that the signal shown in Figure 3-3 is

Communications Engineering. R.C.T. Lee, Mao-Ching Chiu and Jung-Shan Lin
© 2007 John Wiley & Sons (Asia) Pte Ltd

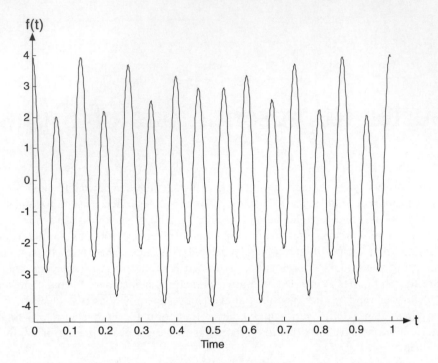

Figure 3-1 A signal

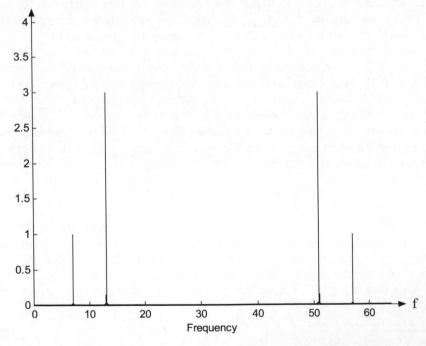

Figure 3-2 The discrete Fourier transform spectrum of the signal in Fig. 3-1 after sampling

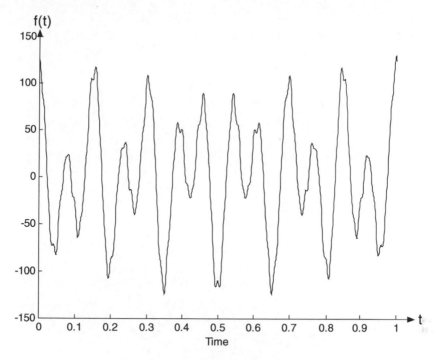

Figure 3-3   A signal with some noise

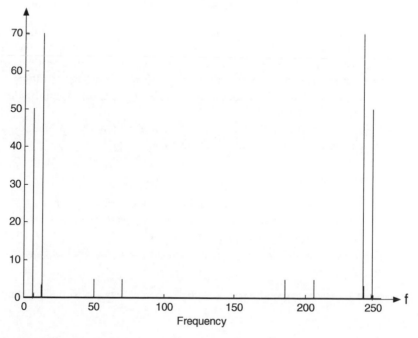

Figure 3-4   The discrete Fourier transform of the signal in Fig. 3-3 after sampling

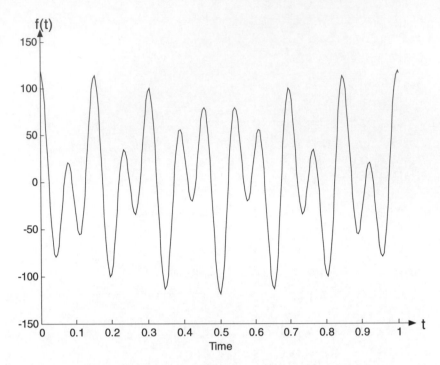

Figure 3-5   The signal obtained by filtering out the noise

composed of two signals with large amplitudes and two signals with small amplitudes. Intuitively, these signals with small amplitudes can be viewed as *noise*. Suppose we filter out these noises and keep only the signals with large amplitudes. We may apply a so-called *inverse discrete Fourier transform* only to these large amplitudes. The result is shown in Figure 3-5. We can see that the result obtained by ignoring the small amplitude signals, which correspond to noise, is all right.

We may say that Fourier transform is, in general, a tool which we can use to analyze a signal to see which frequency components it contains.

3.1 The Fourier Series

A Fourier series is used for representing a continuous-time periodic signal as a weighted superposition of sinusoids. Before starting with the derivations of Fourier series, we should introduce some useful concepts about signals in advance.

3.1.1 Periodic Signals

A continuous-time signal $x(t)$ is said to be **periodic** if there exists a positive constant T such that:

$$x(t) = x(t + T) \quad \text{for all } t, \tag{3-2}$$

where T is the period of the signal. Any signal $x(t)$ for which there is no value of T to satisfy Equation (3-2) is called an **aperiodic** or **nonperiodic** signal. Periodic signals have the following characteristics:

1. A signal $x(t)$ that is periodic with periodic T is also periodic with period nT where n is any positive integer. That is,

$$x(t) = x(t + T) = x(t + nT).$$

2. We define the **fundamental period** T_0 as the smallest value of the period T that satisfies Equation (3-2). The **periodic frequency** f_0 is defined as $f_0 = 1/T_0$. If no ambiguity arises, we may use T, instead of T_0. That is, it is understood that $f_0 = 1/T$.
3. For the units of T in seconds, the units of f are hertz (Hz), which are cycles per second.

Example 3-1
The signal $x(t) = \cos(2\pi f_0 t)$ is periodic, since Equation (3-2) is satisfied. That is, for any $T > 0$:

$$\cos(2\pi f_0(t + T)) = \cos(2\pi f_0 t + 2\pi f_0 T) = \cos(2\pi f_0 t + 2\pi) = \cos(2\pi f_0 t).$$

Clearly, for any integer n, we also have:

$$\cos(2\pi f_0(t + nT)) = \cos(2\pi f_0 t + n2\pi f_0 T) = \cos(2\pi f_0 t + 2n\pi) = \cos(2\pi f_0 t).$$

Example 3-2
Figure 3-6(a) and 3-6(b) present examples of periodic and aperiodic signals, respectively. The periodic signal shown here represents a square wave of amplitude $A = 1$, and period T, and the aperiodic signal represents a rectangular pulse of amplitude A and duration T_1.

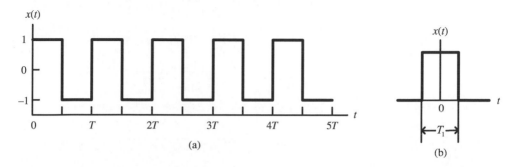

Figure 3-6 Periodic and aperiodic signals: (a) square wave with amplitude $A = 1$, and period T. (b) rectangular pulse of amplitude A and duration T_1.

3.1.2 Orthogonality of Sinusoids

The orthogonality of sinusoids which was discussed in Chapter 2 plays a key role in Fourier representations. In the following, we present the concept again. Two signals are said to be *orthogonal* if their inner product is zero. The *inner product* of two signals $\phi_1(t)$ and $\phi_2(t)$ with period T is defined by the integration of their products over any interval of length T, as given by

$$\langle \phi_1(t), \phi_2(t) \rangle = \int_{\{T\}} \phi_1(t)\phi_2(t)dt \tag{3-3}$$

where the notation $\{T\}$ implies the integration over any interval of length T. Let $\phi_1(t) = \sin(2\pi k f_0 t)$ and $\phi_2(t) = \sin(2\pi n f_0 t)$. Given any nonnegative integers k and n and the fundamental frequency $f_0 = 1/T$, we are interested in the following inner product:

$$
\begin{aligned}
\langle \phi_1(t), \phi_2(t) \rangle &= \int_0^T \sin(2\pi k f_0 t)\sin(2\pi n f_0 t)dt \\
&= \int_0^T \frac{1}{2}[\cos(2\pi(k-n)f_0)t - \cos(2\pi(k+n)f_0)t]dt \\
&= \frac{1}{2}\left[\frac{\sin(2\pi(k-n)f_0)t}{2\pi(k-n)f_0} - \frac{\sin(2\pi(k+n)f_0)t}{2\pi(k+n)f_0}\right]_{t=0}^{t=T=\frac{1}{f_0}} \\
&= 0, \text{ if } k \neq n
\end{aligned} \tag{3-4}
$$

and

$$\langle \phi_1(t), \phi_2(t) \rangle = \frac{1}{2}\int_0^T [1 - \cos(2\pi(k+n)f_0)t]dt = \frac{t}{2}\Big|_{t=0}^{t=T} = T/2 \text{ if } k = n \neq 0.$$

The inner product of $\phi_1(t) = \cos(2\pi k f_0 t)$ and $\phi_2(t) = \cos(2\pi n f_0 t)$ is computed as:

$$
\begin{aligned}
\langle \phi_1(t), \phi_2(t) \rangle &= \int_0^T \cos(2\pi k f_0 t)\cos(2\pi n f_0 t)dt \\
&= \int_0^T \frac{1}{2}[\cos(2\pi(k-n)f_0)t + \cos(2\pi(k+n)f_0)t]dt \\
&= \frac{1}{2}\left[\frac{\sin(2\pi(k-n)f_0)t}{2\pi(k-n)f_0} + \frac{\sin(2\pi(k+n)f_0)t}{2\pi(k+n)f_0}\right]_{t=0}^{t=T=\frac{1}{f_0}} \\
&= 0, \text{ if } k \neq n
\end{aligned} \tag{3-5}
$$

and

$$\langle \phi_1(t), \phi_2(t) \rangle = \frac{1}{2}\int_o^T [1 + \cos(2\pi(k+n)f_0)t]dt = \frac{t}{2}\Big|_{t=0}^{t=T} = T/2 \text{ if } k = n \neq 0.$$

The inner product of $\phi_1(t) = \sin(2\pi k f_0 t)$ and $\phi_2(t) = \cos(2\pi n f_0 t)$ is computed as:

$$\begin{aligned}
\langle \phi_1(t), \phi_2(t) \rangle &= \int_0^T \sin(2\pi k f_0 t) \cos(2\pi n f_0 t) dt \\
&= \int_0^T \frac{1}{2} [\sin(2\pi(k-n)f_0 t) + \sin(2\pi(k+n)f_0 t)] dt \\
&= -\frac{1}{2} \left[\frac{\cos(2\pi(k-n)f_0 t)}{2\pi(k-n)f_0} + \frac{\cos(2\pi(k+n)f_0 t)}{2\pi(k+n)f_0} \right]_{t=0}^{t=T=\frac{1}{f_0}} \\
&= 0, \text{ if } k \neq n
\end{aligned}$$

(3-6)

and

$$\langle \phi_1(t), \phi_2(t) \rangle = \frac{1}{2} \int_0^T [0 + \sin(2\pi(k+n)f_0)t] dt = 0 \text{ if } k = n.$$

Let us summarize the above discussion as follows:

$$\begin{aligned}
&\langle \sin(2\pi k f_c t), \sin(2\pi n f_c t) \rangle \\
&= \langle \cos(2\pi k f_c t), \cos(2\pi n f_c t) \rangle \quad \text{if } k \neq n \\
&= 0
\end{aligned}$$

(3-7)

$$\begin{aligned}
&\langle \sin(2\pi k f_c t), \sin(2\pi n f_c t) \rangle \\
&- \langle \cos(2\pi k f_c t), \cos(2\pi n f_c t) \rangle \quad \text{if } k = n \neq 0 \\
&- \frac{T}{2}
\end{aligned}$$

(3-8)

$$\langle \sin(2\pi k f_c t), \cos(2\pi n f_c t) \rangle = 0 \text{ for all } k \text{ and } n.$$

(3-9)

3.1.3 Derivation of Fourier Series

We begin the derivation of the Fourier series by approximating a periodic signal $x(t)$ having fundamental period $T = 1/f_0$ using the following series:

$$\hat{x}(t) = a_0 + \sum_{k=1}^{\infty} a_k \cos(2\pi k f_0 t) + \sum_{k=1}^{\infty} b_k \sin(2\pi k f_0 t)$$

(3-10)

where a_0, a_k and b_k are constant coefficients to be found later. The series in (3-10) is a linear combination of sine and cosine functions with various frequencies, which are integer multiples of the fundamental frequency f_0 and the coefficient a_0 can be considered the portion with zero frequency (i.e. the DC component) in a signal. We use the orthogonality property in Equations (3-7), (3-8) and (3-9) to find the undefined coefficients. The reader is encouraged to go back to re-read Sections 2.2 and 2.3. Because of the orthogonality of cosine and sine functions, a_k and b_k can be found by multiplying $\hat{x}(t)$ with cosine and sine functions respectively, as suggested in Sections 2.2 and 2.3.

We begin by assuming that we can find the coefficients a_0, a_k and b_k so that $x(t) = \hat{x}(t)$. Then we have

$$\int_{\{T\}} x(t)\phi(t)dt = \int_{\{T\}} \hat{x}(t)\phi(t)dt \qquad (3\text{-}11)$$

where $\phi(t)$ is chosen as follows.

(1) $\phi(t) = 1$ is selected for the computation of a_0. From (3-10) and (3-11), we obtain

$$\int_{\{T\}} x(t)dt$$

$$= \int_{\{T\}} \hat{x}(t)dt$$

$$= \int_0^T a_0 dt + \sum_{k=1}^{\infty} a_k \int_0^T \cos(2\pi k f_0 t)dt + \sum_{k=1}^{\infty} b_k \int_0^T \sin(2\pi k f_0 t)dt. \qquad (3\text{-}12)$$

Each of the integrals in the summation signs in Equation (3-12) is over an integer multiple of the period of the sine or cosine function. Hence they are all zeros. We are left with:

$$\int_{\{T\}} x(t)dt = \int_0^T a_0 dt = a_0 T$$

which implies

$$a_0 = \frac{1}{T} \int_{\{T\}} x(t)dt. \qquad (3\text{-}13)$$

The coefficient a_0 can be viewed as the average value of the signal $x(t)$ over one period.
(2) $\phi(t) = \cos(2\pi n f_0 t)$ with any positive integer n is selected for the computation of a_k. From (3-10) and (3-11) we have:

$$\int_{\{T\}} x(t)\cos(2\pi n f_0 t)dt = \int_{\{T\}} \hat{x}(t)\cos(2\pi n f_0 t)dt$$

$$= a_0 \int_0^T \cos(2\pi n f_0 t)dt + \sum_{k=1}^{\infty} a_k \int_0^T \cos(2\pi k f_0 t)\cos(2\pi n f_0 t)dt$$

$$+ \sum_{k=1}^{\infty} b_k \int_0^T \sin(2\pi k f_0 t)\cos(2\pi n f_0 t)dt.$$

$$(3\text{-}14)$$

The first integral in the right-hand side of Equation (3-14) is zero because we are integrating over an integer multiple of the period. Next note that all of the integrals in the last summation are zero because of Equation (3-9). Finally, all of the integrals in the middle summation vanish except the term with $k = n$, which is equal to $T/2$ because of Equations (3-7) and (3-8). We are thus left with

$$\int_{\{T\}} x(t) \cos(2\pi n f_0 t) dt = a_n \frac{T}{2}.$$

Therefore we arrive at the expression for a_k, which is

$$a_k = \frac{2}{T} \int_{\{T\}} x(t) \cos(2\pi k f_0 t) dt \qquad (3\text{-}15)$$

for any positive integer k.

(3) $\phi(t) = \sin(2\pi n f_0 t)$ with any positive integer n is selected for the computation of b_k. From Equations (3-10) and (3-11), we have

$$\int_{\{T\}} x(t) \sin(2\pi n f_0 t) dt = \int_{\{T\}} \hat{x}(t) \sin(2\pi n f_0 t) dt$$

$$= a_0 \int_0^T \sin(2\pi n f_0 t) dt + \sum_{k=1}^{\infty} a_k \int_0^T \cos(2\pi k f_0 t) \sin(2\pi n f_0 t) dt \qquad (3\text{-}16)$$

$$+ \sum_{k=1}^{\infty} b_k \int_0^T \sin(2\pi k f_0 t) \sin(2\pi n f_0 t) dt.$$

This time the only nonzero integral, which is equal to $T/2$, is in the last summation of Equation (3-16) when $k = n$ due to Equations (3-7), (3-8) and (3-9). Thus, we have

$$\int_{\{T\}} x(t) \sin(2\pi n f_0 t) dt = b_n \frac{T}{2}$$

which implies

$$b_k = \frac{2}{T} \int_{\{T\}} x(t) \sin(2\pi k f_0 t) dt \qquad (3\text{-}17)$$

for any positive integer k.

We conclude that if the signal $x(t)$ can be completely expressed by the estimated signal $\hat{x}(t)$ in Equation (3-10) 'that is, $x(t) = \hat{x}(t)$' then the coefficients a_0, a_k and b_k appearing in Equation (3-10) are given for the formulas in Equations (3-13), (3-15) and (3-17), respectively. These results are summarized as follows.

Definition I

Suppose that $x(t)$, $-\infty < t < \infty$, is a periodic signal with fundamental period $T = 1/f_0$, where f_0 is the fundamental frequency. If there exists a convergent series of the form

$$x(t) = a_0 + \sum_{k=1}^{\infty} a_k \cos(2\pi k f_0 t) + \sum_{k=1}^{\infty} b_k \sin(2\pi k f_0 t) \qquad (3\text{-}18)$$

whose coefficients are given by

$$a_0 = \frac{1}{T} \int_{\{T\}} x(t)dt$$

$$a_k = \frac{2}{T} \int_{\{T\}} x(t)\cos(2\pi k f_0 t)dt \qquad (3\text{-}19)$$

$$b_k = \frac{2}{T} \int_{\{T\}} x(t)\sin(2\pi k f_0 t)dt$$

then the series is called a *Fourier series representation* of the signal $x(t)$. In general, the Fourier series in this form is classified as a **trigonometric Fourier series**.

Example 3-3

To determine the Fourier series representation for the square wave shown in Figure 3-7, we integrate over the period $t = -T/2$ to $t = T/2$ to obtain the Fourier series coefficients. The Fourier series coefficient a_0 is computed as:

$$a_0 = \frac{1}{T} \int_{-T/2}^{T/2} x(t)dt = \frac{1}{T} \int_{-T_s}^{T_s} dt = \frac{2T_s}{T}. \qquad (3\text{-}20)$$

The Fourier series coefficient a_k is computed as:

$$
\begin{aligned}
a_k &= \frac{2}{T} \int_{-T/2}^{T/2} x(t)\cos(2\pi k f_0 t)dt \\
&= \frac{2}{T} \int_{-T_s}^{T_s} \cos(2\pi k f_0 t)dt \\
&= \frac{2}{2\pi k f_0 T} \sin(2\pi k f_0 t)\big|_{-T_s}^{T_s} \qquad (3\text{-}21) \\
&= \frac{2}{\pi k f_0 T} \sin(2\pi k f_0 T_s) \\
&= \frac{2}{\pi k} \sin(2\pi k f_0 T_s).
\end{aligned}
$$

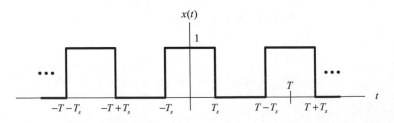

Figure 3-7   A periodic square wave

The Fourier series coefficient b_k is computed as

$$
\begin{aligned}
b_k &= \frac{2}{T} \int_{-T/2}^{T/2} x(t)\sin(2\pi k f_0 t)\,dt \\
&= \frac{2}{T} \int_{-T_s}^{T_s} \sin(2\pi k f_0 t)\,dt \\
&= -\frac{2}{2\pi k f_0 T} \cos(2\pi k f_0 t)\big|_{-T_s}^{T_s} \\
&= 0.
\end{aligned}
\tag{3-22}
$$

Hence the Fourier series representation of the periodic signal $x(t)$ shown in Figure 3-7 is

$$
x(t) = \frac{2T_s}{T} + \sum_{k=1}^{\infty} \frac{2\sin(2\pi k f_0 T_s)}{\pi k} \cos(2\pi k f_0 t).
\tag{3-23}
$$

Since $f_0 = 1/T$, we can express the above equation as the following:

$$
x(t) = \frac{2T_s}{T} + \sum_{k=1}^{\infty} \frac{2\sin(2\pi k T_s/T)}{\pi k} \cos\left(\frac{2\pi k t}{T}\right).
\tag{3-24}
$$

Equations (3-23) and (3-24) indicate that the square wave function presented in Figure 3-7 consists of a series of cosine functions with frequencies $f_0, 2f_0, \ldots$, where $f_0 = 1/T$. The amplitude of the cosine function with frequency $k f_0$ is $\frac{2\sin(2\pi k f_0 T_s)}{\pi k}$. There is also a DC component which is equal to $2T_s/T$.

The value of T_s/T is quite interesting. Consider the special case where $2T_s/T = 1$. In this case, according to Equation (3-24), the coefficient of each cosine function will become 0 and the function contains only the DC component. Why is that? Note that the square wave function in Figure 3-7 becomes 1, a constant, if $2T_s/T = 1$. Consequently, all of the sinusoidal components vanish.

What is the meaning of the Fourier series expressed in Equation (3-23) or Equation (3-24)? They mean that a periodic function can be approximated by an infinite series of cosine functions. In this series, k should be as large as possible. Figure 3-8 shows this idea. As shown in the figure, a small k does not approximate the function very well. But, as k is set to be as large as 1000, the approximated function is almost the same as that shown in Figure 3-7. The power of Fourier series is clearly shown in this experimental result.

The Fourier series analysis shows that the square wave function presented in Figure 3-7 consists only of cosine functions. No sine functions appear in Equation (3-23). Why is so? Taking a look at Figure 3-7, one would note that this square wave function has the following property: it is symmetrical with respect to $t = 0$. That is, $f(-t) = f(t)$ for every t. Any function with the above property is an **even** function. A function is called an **odd** function if $f(-t) = -f(t)$. Since $\cos(-t) = \cos(t)$ for all t, the cosine function is an even function. The sine function is not even. This is why the square wave function illustrated in Figure 3-7 consists of only cosine functions.

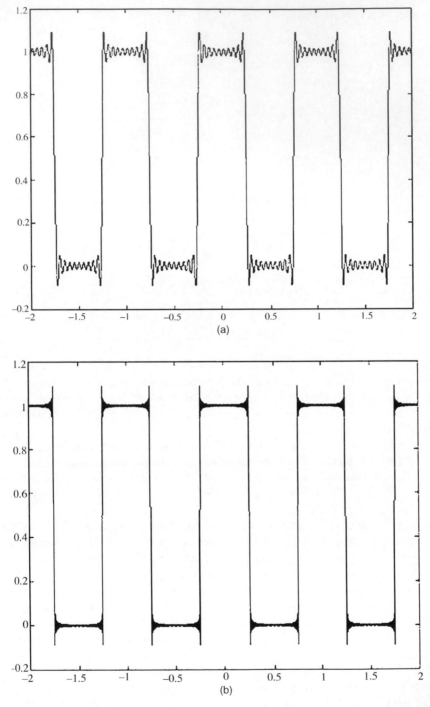

Figure 3-8   The power of Fourier series: (a) $k = 0$ to 10, (b) $k = 0$ to 100, (c) $k = 0$ to 1000

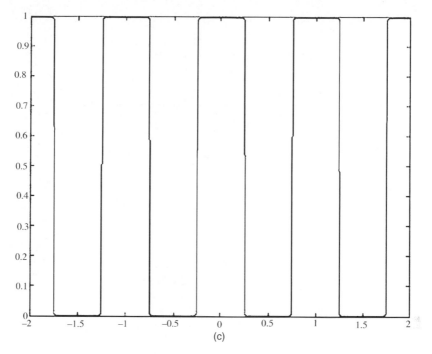

Figure 3-8 *(Continued)*

Example 3-4

Let us now consider the square wave function in Figure 3-9. This function is an odd function. Since sine functions are odd, we expect that this function would consist of only sine functions. This is indeed true. By applying Fourier series analysis to this function, we will find that the Fourier series representation of the function in Figure 3-9 is as follows:

$$
a_0 = \frac{1}{T} \int_{-T/2}^{0} (-1)dt + \frac{1}{T} \int_{0}^{T/2} (1)dt = \frac{1}{T}(-t)\Big|_{-T/2}^{0} + \frac{1}{T}(t)\Big|_{0}^{T/2}
$$

$$
= \frac{1}{T}\left(-\frac{T}{2}\right) + \frac{1}{T}\left(\frac{T}{2}\right) = 0. \tag{3-25}
$$

$$
a_k = \frac{2}{T} \int_{-\frac{T}{2}}^{0} (-1)\cos(2\pi k f_0 t)dt + \frac{2}{T} \int_{0}^{\frac{T}{2}} (1)\cos(2\pi k f_0 t)dt
$$

$$
= \frac{-2}{2\pi k f_0 T}\sin(2\pi k f_0 t)\Big|_{-\frac{T}{2}}^{0} + \frac{2}{2\pi k f_0 T}\sin(2\pi k f_0 t)\Big|_{0}^{\frac{T}{2}}
$$

$$
= \frac{-2}{2\pi k}\sin\left(2\pi k f_0 \frac{T}{2}\right) + \frac{2}{2\pi k}\sin\left(2\pi k f_0 \frac{T}{2}\right)
$$

$$
= 0. \tag{3-26}
$$

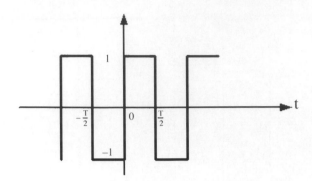

Figure 3-9 An odd function

$$b_k = \frac{2}{T} \int_{-\frac{T}{2}}^{0} (-1)\sin(2\pi k f_0 t)dt + \frac{2}{T} \int_{0}^{\frac{T}{2}} (1)\sin(2\pi k f_0 t)dt$$

$$= \frac{2}{2\pi k} \cos(2\pi k f_0 t)\Big|_{-\frac{T}{2}}^{0} + \frac{-2}{2\pi k}\cos(2\pi k f_0 t)\Big|_{0}^{\frac{T}{2}}$$

$$= \frac{2}{\pi k}\left(1 - \cos\left(2\pi k f_0 \frac{T}{2}\right)\right)$$

$$= \frac{2}{\pi k}(1 - \cos(\pi k f_0 T))$$

$$= \frac{4}{\pi k}\sin^2\left(\frac{\pi k f_0 T}{2}\right). \tag{3-27}$$

Since $f_0 T = 1$, we have

$$x(t) = \sum_{k=1}^{\infty} \frac{4}{\pi k}\sin^2\left(\frac{\pi k}{2}\right)\sin(2\pi k f_0 t). \tag{3-28}$$

In Equation (3-28), no cosine functions appear, as cosine functions are even and the function is an odd function. The reader should also note that there is no DC component in this case. This is because the average value of this function is 0.

Example 3-5
Now let us consider Figure 3-10. In this case, the function is neither even nor odd. Thus, we may expect that both cosine and sine functions would appear. But we can prove that the Fourier series coefficients are as follows (the proof is left as an exercise):

$$a_0 = \frac{1}{2}$$

$$a_k = 0 \tag{3-29}$$

$$b_k = \frac{2}{\pi k}\sin^2\left(\frac{\pi k}{2}\right).$$

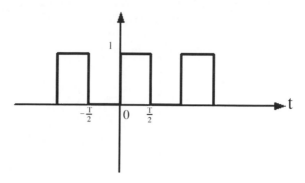

Figure 3-10   A function which is neither even nor odd

Thus, $x(t)$ can be expressed as follows:

$$x(t) = \frac{1}{2} + \sum_{k=1}^{\infty} \frac{2}{\pi k} \sin^2\left(\frac{\pi k}{2}\right) \sin(2\pi k f_0 t). \tag{3-30}$$

The reader may be puzzled about the fact that only sine functions appear in Equation (3-30). Does this imply that the function shown in Figure 3-10 is an odd function? No. It is indeed neither odd nor even. Let us consider Figure 3-10 again. If we subtract 1/2 from the function, the resulting function becomes an odd function and would contain only sine functions. This can also be seen by examining Equation (3-30). There is a DC term, namely 1/2, in it. Consider the function $x'(t) - x(t) - 1/2$.. As indicated in Equation (3-30), $x'(t)$ will only contain sine functions and is an odd function.

Example 3-6
Let us consider Figure 3-11. In this case, the function is really neither even nor odd. In this case, the Fourier series coefficients are found as follows:

$$
\begin{aligned}
a_0 &= \frac{1}{T}\int_{-\frac{T}{2}}^{-\frac{T}{4}}(-1)dt + \frac{1}{T}\int_{0}^{\frac{T}{4}}(1)dt \\[2mm]
&= \frac{1}{T}[-t]_{-T/2}^{-T/4} + \frac{1}{T}[t]_{0}^{T/4} \\[2mm]
&= \frac{1}{T}\left[\frac{T}{4} - \frac{T}{2}\right] + \frac{1}{T}\left[\frac{T}{4}\right] \\[2mm]
&= -\frac{1}{4} + \frac{1}{4} \\[2mm]
&= 0.
\end{aligned}
\tag{3-31}
$$

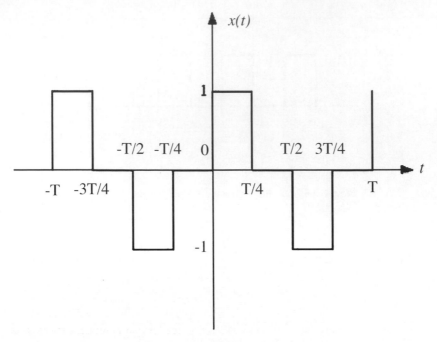

Figure 3-11   Another function which is neither even nor odd

$$a_k = \frac{2}{T}\int_{-T/2}^{-T/4}(-1)\cos(2\pi k f_0 t)dt + \frac{2}{T}\int_{0}^{T/4}(1)\cos(2\pi k f_0 t)dt$$

$$= \frac{2}{T}\left(-\frac{\sin(2\pi k f_0 t)}{2\pi k f_0}\right)_{-\frac{T}{2}}^{-\frac{T}{4}} + \frac{2}{T}\left(\frac{\sin(2\pi k f_0 t)}{2\pi k f_0}\right)_{0}^{\frac{T}{4}} \qquad (3\text{-}32)$$

$$= \frac{-1}{\pi k}\left(\sin\left(\frac{-\pi k}{2}\right) - \sin(-\pi k)\right) + \frac{1}{\pi k}\left(\sin\left(\frac{\pi k}{2}\right) - 0\right)$$

$$= \frac{2}{\pi k}\sin\left(\frac{\pi k}{2}\right).$$

$$b_k = \frac{2}{T}\int_{-T/2}^{-T/4}(-1)\sin(2\pi k f_0 t)dt + \frac{2}{T}\int_{0}^{T/4}(1)\sin(2\pi k f_0 t)dt$$

$$= \frac{2}{T}\left[\frac{\cos(2\pi k f_0 t)}{2\pi k f_0}\right]_{-\frac{T}{2}}^{-\frac{T}{4}} - \frac{2}{T}\left[\frac{\cos(2\pi k f_0 t)}{2\pi k f_0}\right]_{0}^{\frac{T}{4}}$$

$$= \frac{1}{\pi k}\left(\cos\left(\frac{-\pi k}{2}\right) - \cos(-\pi k)\right) - \frac{1}{\pi k}\left(\cos\left(\frac{\pi k}{2}\right) - 1\right) \qquad (3\text{-}33)$$

$$= \frac{1}{\pi k}(1 - \cos(\pi k))$$

$$= \frac{2}{\pi k}\sin^2\left(\frac{\pi k}{2}\right).$$

Thus:

$$x(t) = \sum_{k=1}^{\infty} \frac{2}{\pi k} \sin\left(\frac{\pi k}{2}\right) \cos(2\pi k f_0 t)$$

$$+ \sum_{k=1}^{\infty} \frac{2}{\pi k} \sin^2\left(\frac{\pi k}{2}\right) \sin(2\pi k f_0 t). \tag{3-34}$$

Equation (3-23) indicates that $x(t)$ consists of two series: One is a cosine series and the other is a sine series:

$$x_1(t) = \sum_{k=1}^{\infty} \frac{2}{\pi k} \sin\left(\frac{\pi k}{2}\right) \cos(2\pi k f_0 t) \tag{3-35}$$

$$x_2(t) = \sum_{k=1}^{\infty} \frac{2}{\pi k} \sin^2\left(\frac{\pi k}{2}\right) \sin(2\pi k f_0 t). \tag{3-36}$$

The reader is encouraged to plot $x_1(t)$ and $x_2(t)$ separately. You will find that $x_1(t)$ is an even function which is very similar to that in Figure 3-7 and $x_2(t)$ is an odd function which is the same as that in Figure 3-9. These two functions are illustrated in Figure 3-12. Figure 3-12(a) depicts the even function $x_1(t)$ and 3-12(b) illustrates the odd function $x_2(t)$. Adding these functions up, one gets the function in Figure 3-11 which is neither even nor odd.

3.2 Cosine-only Expansion of Fourier Series

Although the Fourier series may contain both cosine and sine functions, we may still think that the Fourier series contains cosine functions only. This can be explained as follows. For any $k \neq 0$, there are $a_k \cos(2\pi k f_0 t)$ and $b_k \sin(2\pi k f_0 t)$. Let

$$r_k = \sqrt{a_k^2 + b_k^2}. \tag{3-37}$$

Then, we have

$$a_k \cos(2\pi k f_0 t) + b_k \sin(2\pi k f_0 t)$$

$$= r_k \frac{a_k}{r_k} \cos(2\pi k f_0 t) + r_k \frac{b_k}{r_k} \sin(2\pi k f_0 t). \tag{3-38}$$

Let

$$\theta_k = \tan^{-1} \frac{b_k}{a_k}.$$

Then we have

$$\frac{a_k}{r_k} = \cos \theta_k \quad \text{and} \quad \frac{b_k}{r_k} = \sin \theta_k.$$

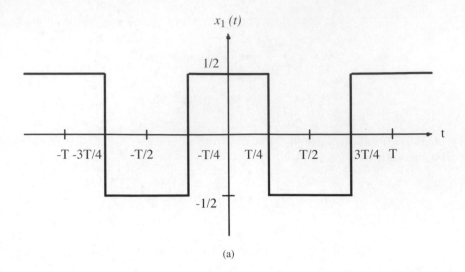

(a)

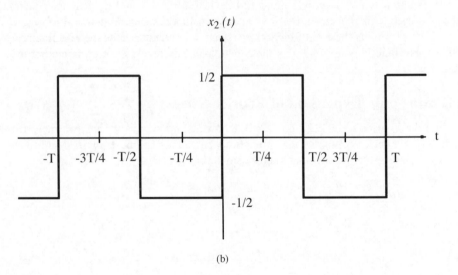

(b)

Figure 3-12   The odd and even functions contained in the function in Figure 3-11: (a) even function $x_1(t)$; (b) odd function $x_2(t)$

We finally have

$$
\begin{aligned}
a_k &\cos(2\pi k f_0 t) + b_k \sin(2\pi k f_0 t) \\
&= r_k \cos(\theta_k) \cos(2\pi k f_0 t) + r_k \sin(\theta_k)\sin(2\pi k f_0 t) \\
&= r_k \cos(2\pi k f_0 t - \theta_k).
\end{aligned}
\tag{3-39}
$$

To get the above equation, Equation A-4 of Appendix A can be used. Equation (3-39) implies that the Fourier series can also be expressed as

$$x(t) = a_0 + \sum_{k=1}^{\infty} r_k \cos(2\pi k f_0 t - \theta_k). \tag{3-40}$$

The term θ_k is a *phase shift*. Thus a periodic function is actually a summation of cosine functions with frequency kf_0 and phase shift θ_k. We may even view the Fourier series representation of a periodic function as a cosine expansion. This concept will be useful for further discussion. There are several special cases related to θ_k.

Case 1: $\theta_k = 0$. This occurs when $b_k = 0$. In this case, there is a cosine function with frequency kf_0 without any phase shift.

Case 2: $\theta_k = \frac{\pi}{2}$. This occurs when $a_k = 0$. In this case, the cosine function with frequency kf_0 becomes a sine function with frequency kf_0 without any phase shift.

Example 3-7
Consider Figure 3-13. In this figure, there are three cases. Figure 3-13(a) shows a square wave function which is an even function. The Fourier series of it therefore contains only cosine functions. By using the result in Example 3-3, we can show that in this case:

$$f_1(t) = \frac{1}{2} + \sum_{k=1}^{\infty} \frac{2}{\pi k} \sin\left(\frac{\pi k}{2}\right) \cos(2\pi k f_c t) \tag{3-41}$$

where $f_c = 1/T$.

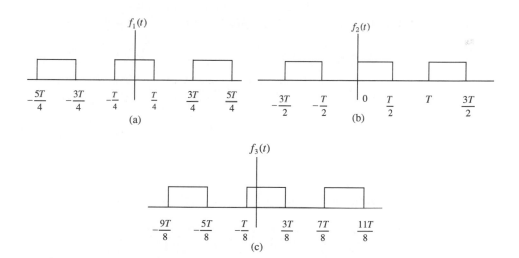

Figure 3-13 The phase shifts: (a) an even function containing only cosine functions; (b)the function with a $\pi/2$ phase shift; (c) the function with a $\pi/4$ phase shift

In Figure 3-13(b), the wave function is shifted to the right and becomes an odd function if we ignore its DC part. This time-shift is 1/4 of the cycle T. In this case, we can show by using the result in Example 3-5 that its Fourier series contains only sine functions as below:

$$f_2(t) = \frac{1}{2} + \sum_{k=1}^{\infty} \frac{2}{\pi k} \sin^2 \frac{\pi k}{2} \sin(2\pi k f_c t). \tag{3-42}$$

Comparing Figures 3-13(a) and 3-13(b), we can see that $f_2(t) = f_1(t - T/4)$. Let us substitute $t - T/4$ for t in Equation (3-41):

$$f_1\left(t - \frac{T}{4}\right) = \frac{1}{2} + \sum_{k=1}^{\infty} \frac{2}{\pi k} \sin\left(\frac{\pi k}{2}\right) \cos\left(2\pi k f_c \left(t - \frac{T}{4}\right)\right)$$

$$= \frac{1}{2} + \sum_{k=1}^{\infty} \frac{2}{\pi k} \sin\left(\frac{\pi k}{2}\right) \left(\cos(2\pi k f_c t) \cos\left(\frac{\pi k}{2}\right) + \sin(2\pi k f_c t)\sin\left(\frac{\pi k}{2}\right)\right)$$

$$= \frac{1}{2} + \sum_{k=1}^{\infty} \frac{2}{\pi k} \left(\sin\left(\frac{\pi k}{2}\right) \cos\left(\frac{\pi k}{2}\right) \cos(2\pi k f_c t) + \sin^2\left(\frac{\pi k}{2}\right) \sin(2\pi k f_c t)\right)$$

$$= \frac{1}{2} + \sum_{k=1}^{\infty} \frac{2}{\pi k} \left(\frac{1}{2}\sin(\pi k) \cos(2\pi k f_c t) + \frac{1}{2}(1 - \cos(\pi k))\sin(2\pi k f_c t)\right)$$

$$= \frac{1}{2} + \sum_{k=1}^{\infty} \frac{1}{\pi k}(1 - \cos(\pi k))\sin(2\pi k f_c t)$$

$$= \frac{1}{2} + \sum_{k=1}^{\infty} \frac{2}{\pi k}\sin^2 \frac{\pi k}{2} \sin(2\pi k f_c t)$$

$$= f_2(t).$$

Now, for the function in Figure 3-13(c), the time-shift is 1/8 of the cycle T. We can show the following by using many equations in Appendix A (the proof is left as an exercise):

$$a_0 = \frac{1}{2} \tag{3-43}$$

$$a_k = \frac{1}{\pi k}\left(\sin\left(\frac{k3\pi}{4}\right) + \sin\left(\frac{k\pi}{4}\right)\right) \tag{3-44}$$

$$b_k = \frac{1}{\pi k}\left(\cos\left(\frac{k3\pi}{4}\right) - \cos\left(\frac{k\pi}{4}\right)\right). \tag{3-45}$$

But

$$\sin\left(\frac{k3\pi}{4}\right) + \sin\left(\frac{k\pi}{4}\right)$$

$$= 2\sin\left(\frac{k}{2}\left(\frac{3\pi}{4} + \frac{\pi}{4}\right)\right) \cos\left(\frac{k}{2}\left(\frac{3\pi}{4} - \frac{\pi}{4}\right)\right) \tag{3-46}$$

$$= 2\sin\left(\frac{k\pi}{2}\right) \cos\left(\frac{k\pi}{4}\right)$$

and

$$\cos\left(\frac{k\pi}{4}\right) - \cos\left(\frac{k3\pi}{4}\right) = 2\sin\left(\frac{k\pi}{2}\right)\sin\left(\frac{k\pi}{4}\right). \tag{3-47}$$

Thus:

$$
\begin{aligned}
f_3(t) &= \frac{1}{2} + \sum_{k=1}^{\infty} \frac{2}{\pi k} \sin\left(\frac{k\pi}{2}\right)\left(\cos\left(\frac{k\pi}{4}\right)\cos(2\pi k f_c t) + \sin\left(\frac{k\pi}{4}\right)\sin(2\pi k f_c t)\right) \\
&= \frac{1}{2} + \sum_{k=1}^{\infty} \frac{2}{\pi k} \sin\left(\frac{k\pi}{2}\right)\cos\left(2\pi k f_c t - \frac{k\pi}{4}\right).
\end{aligned} \tag{3-48}
$$

In the following, we shall prove that $f_3(t) = f_1\left(t - \frac{T}{8}\right)$. From (3-41), we have

$$f_1(t) = \frac{1}{2} + \sum_{k=1}^{\infty} \frac{2}{\pi k} \sin\left(\frac{\pi k}{2}\right)\cos(2\pi k f_c t).$$

Therefore:

$$
\begin{aligned}
f_1\left(t - \frac{T}{8}\right) &= \frac{1}{2} + \sum_{k=1}^{\infty} \frac{2}{\pi k} \sin\left(\frac{\pi k}{2}\right)\cos\left(2\pi k f_c \left(t - \frac{T}{8}\right)\right) \\
&= \frac{1}{2} + \sum_{k=1}^{\infty} \frac{2}{\pi k} \sin\left(\frac{\pi k}{2}\right)\cos\left(2\pi k f_c t - \frac{2\pi k f_c T}{8}\right) \\
&= \frac{1}{2} + \sum_{k=1}^{\infty} \frac{2}{\pi k} \sin\left(\frac{\pi k}{2}\right)\cos\left(2\pi k f_c t - \frac{k\pi}{4}\right).
\end{aligned} \tag{3-49}
$$

Comparing Equations (3-48) and (3-49), we have $f_3(t) = f_1\left(t - \frac{T}{8}\right)$.

3.3 Fourier Series in Complex Exponentials

As we showed at the end of the last section, the Fourier series contains cosine functions. Since both cosine and sine functions are closely related to complex exponentials which we will introduce in this section, a much more compact form of the Fourier series than the trigonometric Fourier series previously discussed consists of a sum of complex exponential time waveforms. Hence, let us review the concept of complex numbers in advance.

As we know, the square of a real number is always nonnegative, and hence the square root of a negative number cannot exist; it is considered to be an 'imaginary' number. The imaginary number corresponding to the square root of a minus one is given the symbol j. That is, $j = \sqrt{-1}$. The complex number z can be expressed in the following two ways.

The **Cartesian** or **rectangular** form for z is:

$$z = x + jy \tag{3-50}$$

where x and y are real numbers referred to respectively as the **real part** and the **imaginary part** of z. The notations $x = \text{Re}\{z\}$ and $y = \text{Im}\{z\}$ are often used.

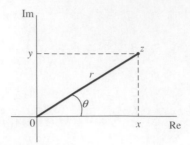

Figure 3-14 The complex plane

The complex number z can also be represented in **polar form** as:

$$z = re^{j\theta} \tag{3-51}$$

where $r \geq 0$ is the **magnitude** of z and θ is the **angle** or **phase** of z. These quantities will often be written as $r = |z|$ and $\theta = \angle z$.

The relationship between these two representations of complex numbers can be determined either from the **Euler's formula**:

$$e^{j\theta} = \cos\theta + j\sin\theta \tag{3-52}$$

or by plotting z in the complex plane, as shown in Figure 3-14, in which the coordinate axes are $\text{Re}\{z\}$ along the horizontal axis and $\text{Im}\{z\}$ along the vertical axis. With respect to this geographical representation, x and y are the rectangular coordinates of z, and r and θ are its polar coordinates. Therefore the relationship for converting from rectangular form to polar coordinates is given by

$$r = \sqrt{x^2 + y^2} \quad \text{and} \quad \theta = \tan^{-1}\frac{y}{x} \tag{3-53}$$

and the relationship for converting from polar to rectangular coordinates is given by

$$x = r\cos\theta \quad \text{and} \quad y = r\sin\theta. \tag{3-54}$$

From the definition of Euler's relation in (3-52), we can obtain

$$e^{-j\theta} = e^{j(-\theta)} = \cos(-\theta) + j\sin(-\theta) = \cos\theta - j\sin\theta, \tag{3-55}$$

and hence we have the following important identities:

$$\cos\theta = \frac{1}{2}(e^{j\theta} + e^{-j\theta}) \text{ and } \sin\theta = \frac{1}{2j}(e^{j\theta} - e^{-j\theta}) \tag{3-56}$$

which express sinusoids as linear combinations of complex exponentials with complex coefficients. We are going to use them for deriving the complex exponential Fourier series later.

Complex Exponential Representation of Fourier Series

In general, we would like to express Fourier series in the complex exponential form because it consists only of complex exponential time functions in the form $e^{j2\pi ft}$, so that the function always retains the same form after differentiation and integration. Looking to express sinusoidal time functions as linear combinations of complex exponential time waveforms with complex coefficients, we can substitute $\theta = k2\pi f_0 t$, where k is any integer, into Equation (3-56) to yield

$$\cos(2\pi k f_0 t) = \frac{1}{2}(e^{j2\pi k f_0 t} + e^{-j2\pi k f_0 t}) \tag{3-57}$$

$$\sin(2\pi k f_0 t) = \frac{1}{2j}(e^{j2\pi k f_0 t} - e^{-j2\pi k f_0 t}). \tag{3-58}$$

Now, we are going to derive the complex exponential Fourier series from the trigonometric Fourier series in Definition I. Given a periodic signal $x(t)$ with fundamental period $T = 1/f_0$, it can be written in the trigonometric Fourier series from Equation (3-10). If Equations (3-57) and (3-58) are substituted into (3-10), we have

$$
\begin{aligned}
x(t) &= a_0 + \sum_{k=1}^{\infty} a_k \cos(2\pi k f_0 t) + \sum_{k=1}^{\infty} b_k \sin(2\pi k f_0 t) \\
&= a_0 + \sum_{k=1}^{\infty} a_k \frac{e^{j2\pi k f_0 t} + e^{-j2\pi k f_0 t}}{2} + \sum_{k=1}^{\infty} b_k \frac{e^{j2\pi k f_0 t} - e^{-j2\pi k f_0 t}}{2j} \\
&= a_0 + \sum_{k=1}^{\infty} \left(\frac{a_k}{2} + \frac{b_k}{2j}\right) e^{j2\pi k f_0 t} + \sum_{k=1}^{\infty} \left(\frac{a_k}{2} - \frac{b_k}{2j}\right) e^{-j2\pi k f_0 t} \\
&= a_0 e^{j2\pi 0 f_0 t} + \sum_{k=1}^{\infty} \left(\frac{a_k}{2} + \frac{b_k}{2j}\right) e^{j2\pi k f_0 t} + \sum_{k=-\infty}^{-1} \left(\frac{a_{-k}}{2} - \frac{b_{-k}}{2j}\right) e^{j2\pi k f_0 t} \\
&= \sum_{k=-\infty}^{\infty} X_k e^{j2\pi k f_0 t}
\end{aligned}
\tag{3-59}
$$

where the coefficients X_k are as follows.

Case 1: $k = 0$. From Equations (3-13) and (3-59):

$$X_0 = a_0 = \frac{1}{T}\int_{\{T\}} x(t)dt. \tag{3-60}$$

Case 2: $k > 0$. From Equations (3-15), (3-17) and (3-59), we have:

$$
\begin{aligned}
X_k &= \frac{a_k}{2} + \frac{b_k}{2j} \\
&= \frac{1}{T} \int_{\{T\}} x(t) \cos(2\pi k f_0 t) dt + \frac{1}{Tj} \int_{\{T\}} x(t) \sin(2\pi k f_0 t) dt \\
&= \frac{1}{T} \int_{\{T\}} x(t) (\cos(2\pi k f_0 t) - j \sin(2\pi k f_0 t)) dt \\
&= \frac{1}{T} \int_{\{T\}} x(t) e^{-j2\pi k f_0 t} dt.
\end{aligned}
\tag{3-61}
$$

Case 3: $k < 0$. Again, from Equations (3-15), (3-17) and (3-59), we have:

$$
\begin{aligned}
X_k &= \frac{a_{-k}}{2} - \frac{b_{-k}}{2j} \\
&= \frac{1}{T} \int_{\{T\}} x(t) \cos 2(\pi k f_0 t) dt - \frac{1}{Tj} \int_{\{T\}} x(t) \sin(-2\pi k f_0) t dt \\
&= \frac{1}{T} \int_{\{T\}} x(t) (\cos(2\pi k f_0 t) - j \sin(2\pi k f_0 t)) dt \\
&= \frac{1}{T} \int_{\{T\}} x(t) e^{-j2\pi k f_0 t} dt.
\end{aligned}
\tag{3-62}
$$

From Equations (3-60), (3-61) and (3-62), we have:

$$
X_k = \frac{1}{T} \int_{\{T\}} x(t) e^{-j2\pi k f_0 t} dt
\tag{3-63}
$$

for any k. Therefore, we can summarize the complex exponential Fourier series in Definition II.

Definition II

Suppose that $x(t), -\infty < t < \infty$, is a periodic signal with fundamental period $T = 1/f_0$, where f_0 is the fundamental frequency. If there exists a convergent series of the form

$$
x(t) = \sum_{k=-\infty}^{\infty} X_k e^{j2\pi k f_0 t}
\tag{3-64}
$$

whose coefficients can be computed from

$$
X_k = \frac{1}{T} \int_{\{T\}} x(t) e^{-j2\pi k f_0 t} dt,
\tag{3-65}
$$

then the series is called a *complex exponential Fourier series* representation of the signal $x(t)$.

The reader should note that $x(t)$ is a real function. In Definition II, the Fourier series contains imaginary part because $e^{j2\pi k f_0 t} = \cos(2\pi k f_0 t) + j \sin(2\pi k f_0 t)$. Suppose that X_k is real. How the imaginary term, namely $j \sin 2\pi k f_0 t$, has no contribution in the final series will be explained. X_k itself can also be complex. This makes things even more puzzling. A formal proof that the imaginary part will finally vanish will be given at the end of this section.

Another point to mention here is that, in Section 3.1, all of the frequencies are positive. In this section, there are now *negative frequencies* because the summation goes through from minus infinity to plus infinity. Every X_{-k} is associated with a negative frequency $-kf_0$. The reader might be quite puzzled at the physical meaning of negative frequencies.

Physically, of course, *there are no negative frequencies*. The existence of negative frequencies in the analysis is due to the fact that the Fourier series actually contains cosine functions as shown in Equation (3-40). In the complex exponential representation of the Fourier series, each positive frequency kf_0 corresponds to the term $e^{j2\pi k f_0 t}$. If only such a positive frequency exists and no negative frequency exists, the Fourier series cannot contain cosine functions. Now, $e^{-j2\pi k f_0 t}$ exists. Note that

$$e^{j2\pi k f_0 t} + e^{-j2\pi k f_0 t} = 2 \cos(2\pi k f_0 t).$$

Thus, we can see that the existence of negative frequencies gives us cosine functions. *We may say that the negative frequencies are not physical entities. They are necessary because $e^{j2\pi k f_0 t}$ must be combined with $e^{-j2\pi k f_0 t}$ to produce $2\cos(2\pi k f_0 t)$. In other words, negative frequencies are only mathematical terms. They exist only because we are using the complex exponential representation for Fourier series.*

Example 3-8
We want to determine the complex exponential Fourier series for the square wave shown in Figure 3-7. We integrate over the period from $t - T/2$ to $t = T/2$ to obtain the Fourier series coefficients with $k \neq 0$:

$$
\begin{aligned}
X_k &= \frac{1}{T} \int_{T/2}^{T/2} x(t) e^{-j2\pi k f_0 t} dt \\
&= \frac{1}{T} \int_{-T_s}^{T_s} e^{-j2\pi k f_0 t} dt \\
&= -\frac{1}{j2\pi k f_0 T} e^{-j2\pi k f_0 t} \Big|_{-T_s}^{T_s} \\
&= \frac{2}{2\pi k} \left(\frac{e^{j2\pi k f_0 T_s} - e^{-j2\pi k f_0 T_s}}{2j} \right) \\
&= \frac{1}{\pi k} \sin(2\pi k f_0 T_s).
\end{aligned}
\tag{3-66}
$$

For $k = 0$, we have

$$X_0 = \frac{1}{T} \int_{-T_s}^{T_s} dt = \frac{2T_s}{T}. \tag{3-67}$$

Using L'Hôpital's rule, it is straightforward to show that

$$\lim_{k\to 0}\frac{1}{\pi k}\sin(2\pi k f_0 T_s) = \lim_{k\to 0} 2\frac{T_s}{T}\cos(2\pi k f_0 T_s) = \frac{2T_s}{T}.$$

Therefore, we write the Fourier series coefficients of $x(t)$ for all k as

$$X_k = \frac{1}{\pi k}\sin(2\pi k f_0 T_s),$$

with the understanding that X_0 is obtained as a limit. In this problem, X_k is real-valued. Substituting $f_0 = 1/T$ gives X_k a function of the ratio T_s/T, as shown in the following:

$$X_k = \frac{1}{k\pi}\sin\left(\frac{2\pi k T_s}{T}\right) = X_{-k}.$$

Through Equation (3-64), we have

$$x(t) = \sum_{k=-\infty}^{\infty} X_k e^{j2\pi k f_0 t}$$

$$= \sum_{k=-\infty}^{\infty} X_k(\cos(2\pi k f_0 t) + j\sin(2\pi k f_0 t))$$

$$= X_0 + \sum_{k=1}^{\infty} X_{-k}(\cos(2\pi(-k)f_0 t) + j\sin(2\pi(-k)f_0 t))$$

$$+ \sum_{k=1}^{\infty} X_k(\cos(2\pi k f_0 t) + j\sin(2\pi k f_0 t))$$

$$= X_0 + \sum_{k=1}^{\infty} X_k(\cos(2\pi k f_0 t) - j\sin(2\pi k f_0 t))$$

$$+ \sum_{k=1}^{\infty} X_k(\cos(2\pi k f_0 t) + j\sin(2\pi k f_0 t))$$

$$= X_0 + \sum_{k=1}^{\infty} 2X_k \cos(2\pi k f_0 t)$$

$$= \frac{2T_s}{T} + \sum_{k=1}^{\infty} \frac{2}{\pi k}\sin\left(\frac{2\pi k T_s}{T}\right)\cos(2\pi k f_0 t) \qquad (3\text{-}68)$$

Figure 3-15 depicts $X_k, -50 \le k \le 50$ for $T_s/T = 1/4$ and $T_s/T = 1/16$. Examine Figure 3-15(b). The first point in which the function X_k intersects the frequency axis is called the *null point*. If the null point is far away from 0, it means that the Fourier series contains many high-frequency signals. *As shown in Figure 3-15, if the width of the pulse T_s is small, the Fourier series contains higher frequency signals. This means that we have a train of pulses and all of the pulses have rather small width, we have to cope with a large number of high-frequency signals.* This concept is quite important.

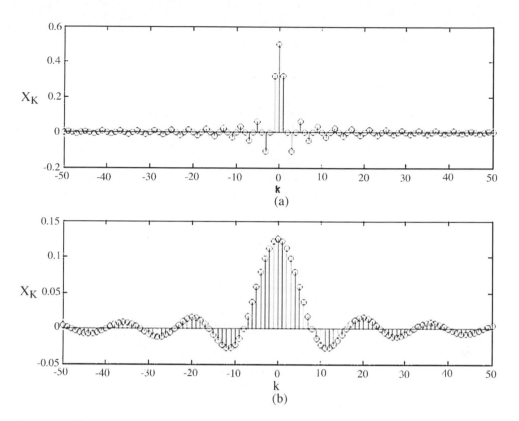

Figure 3-15 The Fourier series coefficients, $X_k, -50 \leq k \leq 50$, for two square waves: (a) $T_s/T - 1/4$ and (b) $T_s/T = 1/16$

It is appropriate to examine the result of expressing the Fourier series in complex exponential form. From Equation (3-64) in Definition II, we know that

$$x(t) = \sum_{k=-\infty}^{\infty} X_k e^{j2\pi k f_0 t}.$$

As we noted before, $x(t)$ is real in this example and X_k is also real as shown in Equation (3-66). *How can the series contain imaginary terms*, *namely* $jX_k \sin(2\pi k f_0 t)$? This question will now be answered. First, let us examine X_k.

Since

$$X_k = \frac{1}{k\pi} \sin\left(\frac{2\pi k T_s}{T}\right), \tag{3-69}$$

we have

$$X_{-k} = \frac{1}{-k\pi} \sin\left(-\frac{2\pi k T_s}{T}\right) = X_k. \tag{3-70}$$

Now consider

$$X_{-k}e^{-j2\pi kf_0 t} + X_k e^{j2\pi kf_0 t}$$
$$= X_k(\cos(2\pi kf_0 t) - j\sin(2\pi kf_0 t)) + X_k(\cos(2\pi kf_0 t) + j\sin(2\pi kf_0 t)) \qquad (3\text{-}71)$$
$$= 2X_k \cos(2\pi kf_0 t).$$

Consequently, in the Fourier series, every sine function disappears because $X_{-k} = X_k$ in this case and $\sin(-\theta) = -\sin(\theta)$. Thus, the complex exponential form Fourier series actually contains no imaginary parts. The reader should compare the result in this example with that in Example 3-3. They are exactly the same.

There is actually another way to obtain X_k. Note that $X_k = a_k/2 + b_k/2j$ from Equation (3-59). Using the results of a_k and b_k from Example 3-3, we can obtain exactly the same result.

Example 3-9

Consider the square wave function in Example 3-4. In this case, $X_0 = 0$ and

$$X_k = \frac{1}{T}\int_{-\frac{T}{2}}^{0} (-1)e^{-j2\pi kf_0 t}\,dt + \frac{1}{T}\int_{0}^{\frac{T}{2}} (1)e^{-j2\pi kf_0 t}\,dt$$

$$= \frac{1}{j2\pi kf_0 T}e^{-j2\pi kf_0 t}\Big|_{-\frac{T}{2}}^{0} + \frac{-1}{j2\pi kf_0 T}e^{-j2\pi kf_0 t}\Big|_{0}^{\frac{T}{2}}$$

$$= \frac{1}{j2\pi k}\left(1 - e^{j2\pi kf_0\frac{T}{2}}\right) - \frac{1}{j2\pi k}\left(e^{-j2\pi kf_0\frac{T}{2}} - 1\right)$$

$$= \frac{1}{j2\pi k}\left(1 - e^{j2\pi kf_0\frac{T}{2}} + 1 - e^{-j2\pi kf_0\frac{T}{2}}\right)$$

$$= \frac{1}{j\pi k}(1 - \cos \pi k)$$

$$= \frac{2}{j\pi k}\sin^2\left(\frac{\pi k}{2}\right). \qquad (3\text{-}72)$$

Thus we have

$$X_{-k} = -X_k,$$

and

$$X_{-k}e^{-j2\pi kf_0 t} + X_k e^{j2\pi kf_0 t}$$
$$= X_k(e^{j2\pi kf_0 t} - e^{-j2\pi kf_0 t})$$
$$= \frac{2}{j\pi k}\sin^2\left(\frac{\pi k}{2}\right)2j\sin(2\pi kf_0 t)$$
$$= \frac{4}{\pi k}\sin^2\left(\frac{\pi k}{2}\right)\sin(2\pi kf_0 t).$$

Thus:

$$x(t) = \sum_{k=1}^{\infty} \frac{4}{\pi k}\sin^2\left(\frac{\pi k}{2}\right)\sin(2\pi kf_0 t). \qquad (3\text{-}73)$$

Equation (3-73) is the same as Equation (3-28) in Example 3-4. We may in fact obtain Equation (3-72) directly from results of a_k and b_k in Example (3-4) by using $x_k = a_k/2 + b_k/2j$.

Equation (3-72) shows that, in this case, X_k contains only an imaginary part. It will be shown later that, whenever this happens, the Fourier series contains only sine functions.

Example 3-10
Consider the function in Example 3-5. We can prove that in this case $X_0 = 1/2$ and:

$$X_k = \frac{1}{j2\pi k}(1 - e^{-j2\pi k f_0 T/2}) = \frac{1}{j2\pi k}(1 - \cos(\pi k)) = \frac{1}{j\pi k}\sin^2\left(\frac{\pi k}{2}\right)$$

$$X_{-k} = \frac{1}{j2\pi k}(e^{j2\pi k f_0 T/2} - 1) = \frac{-1}{j2\pi k}(1 - \cos(\pi k)) = \frac{-1}{j\pi k}\sin^2\left(\frac{\pi k}{2}\right).$$

(3-74)

Thus $X_{-k} = -X_k$. Therefore we have

$$X_{-k}e^{-j2\pi k f_0 t} + X_k e^{j2\pi k f_0 t}$$
$$= X_k(e^{j2\pi k f_0 t} - e^{-j2\pi k f_0 t})$$
$$= X_k(2j)\sin(2\pi k f_0 t)$$
$$= \frac{1}{j2\pi k}(1 - \cos(\pi k))(2j)\sin(2\pi k f_0 t)$$
$$= \frac{1}{\pi k}(1 - \cos(\pi k))\sin(2\pi k f_0 t)$$
$$= \frac{2}{\pi k}\sin^2\left(\frac{\pi k}{2}\right)\sin(2\pi k f_0 t).$$

Thus:

$$x(t) = \frac{1}{2} + \sum_{k=1}^{\infty}\frac{2}{\pi k}\sin^2\left(\frac{\pi k}{2}\right)\sin(2\pi k f_0 t).$$

(3-75)

Equation (3-75) is the same as Equation (3-30) in Example 3-5.

Example 3-11
Consider the function in Example 3-6. In this case, we have

$$X_0 = 0$$

(3-76)

$$X_k = \frac{1}{T}\int_{-T/2}^{-T/4}(-1)e^{-j2\pi k f_0 t}dt + \frac{1}{T}\int_{0}^{T/4}(1)e^{-j2\pi k f_0 t}dt$$

$$= \frac{1}{T}\left[\frac{1}{j2\pi k f_0}e^{-j2\pi k f_0 t}\right]_{-\frac{T}{2}}^{-\frac{T}{4}} - \frac{1}{T}\left[\frac{1}{j2\pi k f_0}e^{-j2\pi k f_0 t}\right]_{0}^{\frac{T}{4}}$$

$$= \frac{1}{j2\pi k}\left(e^{\frac{j\pi k}{2}} - e^{j\pi k}\right) - \frac{1}{j2\pi k}\left(e^{-\frac{j\pi k}{2}} - 1\right)$$

$$= \frac{1}{j2\pi k}\left(e^{\frac{j\pi k}{2}} - e^{-\frac{j\pi k}{2}} + 1 - e^{j\pi k}\right)$$

$$= \frac{1}{j2\pi k}\left(2j\sin\left(\frac{\pi k}{2}\right) + 1 - \cos(\pi k)\right)$$

$$= \frac{1}{\pi k}\sin\left(\frac{\pi k}{2}\right) + \frac{1}{j2\pi k}(1 - \cos(\pi k))$$

$$= \frac{1}{\pi k}\sin\left(\frac{\pi k}{2}\right) - j\frac{1}{2\pi k}(1 - \cos(\pi k))$$

$$= \frac{1}{\pi k}\sin\left(\frac{\pi k}{2}\right) - j\frac{1}{\pi k}\sin^2\left(\frac{\pi k}{2}\right) \tag{3-77}$$

and

$$X_{-k} = -\frac{1}{\pi k}\sin\left(\frac{-\pi k}{2}\right) + j\frac{1}{\pi k}\sin^2\left(\frac{-\pi k}{2}\right)$$

$$= \frac{1}{\pi k}\sin\left(\frac{\pi k}{2}\right) + j\frac{1}{\pi k}\sin^2\left(\frac{\pi k}{2}\right). \tag{3-78}$$

Let:

$$A = \frac{1}{\pi k}\sin\left(\frac{\pi k}{2}\right) \text{ and } B = -\frac{1}{\pi k}\sin^2\left(\frac{\pi k}{2}\right).$$

Then:

$$X_k = A + jB = re^{j\alpha}$$
$$X_{-k} = A - jB = re^{-j\alpha}$$

where $r = \sqrt{A^2 + B^2}$ and $\alpha = \tan^{-1}\frac{B}{A}$.

$$X_{-k}e^{-j2\pi k f_0 t} + X_k e^{j2\pi k f_0 t}$$

$$= re^{-j\alpha}e^{-j2\pi k f_0 t} + re^{j\alpha}e^{j2\pi k f_0 t}$$

$$= r(e^{-j(2\pi k f_0 t + \alpha)} + e^{j(2\pi k f_0 t + \alpha)}) \tag{3-79}$$

$$= 2r\cos(2\pi k f_0 t + \alpha)$$

$$= 2r\cos\alpha\cos(2\pi k f_0 t) - 2r\sin\alpha\sin(2\pi k f_0 t).$$

But

$$r\cos\alpha = A = \frac{1}{\pi k}\sin\left(\frac{\pi k}{2}\right) \tag{3-80}$$

and

$$r \sin\alpha = B = \frac{-1}{\pi k} \sin^2\left(\frac{\pi k}{2}\right).$$

(3-81)

Thus:

$$X_{-k}e^{-j2\pi k f_0 t} + X_k e^{j2\pi k f_0 t}$$

$$= \frac{2}{\pi k} \sin\left(\frac{\pi k}{2}\right) \cos(2\pi k f_0 t) + \frac{2}{\pi k} \sin^2\left(\frac{\pi k}{2}\right) \sin(2\pi k f_0 t).$$

Finally, we have

$$x(t) = \sum_{k=1}^{\infty} \frac{2}{\pi k} \sin\left(\frac{\pi k}{2}\right) \cos(2\pi k f_0 t)$$

$$+ \sum_{k=1}^{\infty} \frac{2}{\pi k} \sin^2\left(\frac{\pi k}{2}\right) \sin(2\pi k f_0 t).$$

(3-82)

Comparing Equations (3-82) and (3-34), we note that in this case, the complex Fourier series representation is the same as the trigonometric Fourier series representation. In general, note that

$$a + jb = \sqrt{a^2 + b^2}\left(\frac{a}{\sqrt{a^2 + b^2}} + j\frac{b}{\sqrt{a^2 + b^2}}\right)$$

$$= r(\cos\theta + j\sin\theta)$$

where $r = \sqrt{a^2 + b^2}$ and $\theta = \tan^{-1}\left(\frac{b}{a}\right)$.

Consider Definition II again. Using Equation (3-59), we can see that for $k > 0$:

$$X_k = \frac{1}{2}(a_k - jb_k) \quad \text{and} \quad X_{-k} = \frac{1}{2}(a_{-k} - jb_{-k}).$$

But, we can easily prove that

$$a_k = a_{-k} \quad \text{and} \quad b_k = -b_{-k}.$$

Thus, we have

$$X_k = \frac{1}{2}(a_k - jb_k) = \frac{1}{2}r_k e^{-j\theta_k} \quad \text{and} \quad X_{-k} = \frac{1}{2}(a_k + jb_k) = \frac{1}{2}r_k e^{j\theta_k},$$

where $r_k = \sqrt{a_k^2 + b_k^2}$ and $\theta_k = \tan^{-1}\left(\frac{b_k}{a_k}\right)$. Thus:

$$X_{-k}e^{-j2\pi k f_0 t} + X_k e^{j2\pi k f_0 t} = \frac{1}{2}(r_k e^{j\theta_k}e^{-j2\pi k f_0 t} + r_k e^{-j\theta_k}e^{j2\pi k f_0 t})$$

$$= \frac{1}{2}r_k(e^{-j(2\pi k f_0 t - \theta_k)} + e^{j(2\pi k f_0 t - \theta_k)})$$

(3-83)

$$= r_k \cos(2\pi k f_0 t - \theta_k).$$

From Equation (3-64), we can have the following:

$$x(t) = x_0 + \sum_{k=1}^{\infty} r_k \cos(2\pi k f_0 t - \theta_k).$$ (3-84)

From this we can see that Equation (3-84) is identical to Equation (3-40). This is why, although the complex exponential form Fourier series contains imaginary terms, the imaginary terms will finally disappear because X_{-k} is a conjugate of X_k.

Again, as in the case where the Fourier series is in trigonometric form, there are special cases for the exponential form.

Case 1: $\theta_k = 0$. This occurs when $b_k = 0$. In this case, there is a cosine function with frequency kf_0 without any phase shift.

Case 2: $\theta_k = \frac{\pi}{2}$. This occurs when $a_k = 0$. In this case, the cosine function with frequency kf_0 becomes a sine function with frequency kf_0 without any phase shift.

We may claim the following: If X_k contains only a real part, the Fourier series contains cosine functions only. If X_k contains only an imaginary part, the Fourier series contains only sine functions. Here the phase shift is $\pi/2$. If X_k contains both real and imaginary parts, the Fourier series contains both cosine and sine functions, or equivalently, cosine functions with phase shifts.

3.4 The Fourier Transform

In contrast to the Fourier series representation for periodic signals, the *Fourier transform* is used to represent a continuous-time nonperiodic signal as a superposition of complex sinusoids. The idea that a periodic signal can be represented as a sum of sines and cosines with Fourier series is very powerful. We would like to extend this result to aperiodic signals. Extension of the Fourier series to aperiodic signals can be done by extending the period to infinity. In order to take this approach, we assume that the Fourier series of periodic extension of the nonperiodic signal $x(t)$ exists. The nonperiodic signal $x(t)$ is defined in the interval $t_0 \leq t \leq t_0 + \bar{T}$ with $\bar{T} > 0$. That is, $x(t) = 0$ outside of this interval. We can generate the periodic extension $\bar{x}(t)$ of the nonperiodic signal $x(t)$ by choosing a constant T with $T > 2(t_0 + \bar{T})$ for $|t_0 \geq -\frac{\bar{T}}{2}|$ and $T > -2t_0$ for $|t_0 < -\frac{\bar{T}}{2}|$, and denoting

$$\bar{x}(t) = x(t), \text{ with } -\frac{T}{2} + kT \leq t \leq \frac{T}{2} + kT,$$ (3-85)

where k is any nonnegative integer, and clearly T is the fundamental period of $\bar{x}(t)$. Conversely, we can express the nonperiodic signal $x(t)$ in terms of the periodic signal $\bar{x}(t)$ by using

$$x(t) = \begin{cases} \bar{x}(t), & -\frac{T}{2} \leq t \leq \frac{T}{2} \\ 0 & \text{otherwise.} \end{cases}$$ (3-86)

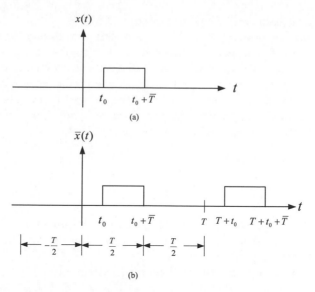

Figure 3-16   The changing of a nonperiodic signal into a periodic signal: (a) $x(t)$ and (b) $\bar{x}(t)$

The above discussion is illustrated in Figure 3-16.

If we let the period T approach infinity, then in the limit, the periodic signal will be reduced to the aperiodic signal. That is:

$$x(t) = \lim_{T \to \infty} \bar{x}(t). \tag{3-87}$$

According to Definition II, this periodic signal $\bar{x}(t)$ with fundamental period $T = 1/f_0$ has a complex exponential Fourier series which is given by

$$\bar{x}(t) = \sum_{k=-\infty}^{\infty} X_k e^{j2\pi k f_0 t} \tag{3-88}$$

where

$$X_k = \frac{1}{T} \int_{-\frac{T}{2}}^{\frac{T}{2}} \bar{x}(t) e^{-j2\pi k f_0 t} dt. \tag{3-89}$$

As far as the integration of Equation (3-89) is concerned, the integrand on this integral can be rewritten as

$$X_k = \frac{1}{T} \int_{-\frac{T}{2}}^{\frac{T}{2}} \bar{x}(t) e^{-j2\pi k f_0 t} dt = \frac{1}{T} \int_{-\infty}^{\infty} x(t) e^{-j2\pi k f_0 t} dt. \tag{3-90}$$

Equation (3-90) holds because of Equations (3-86) and (3-87).

In Equation (3-90), we have kf_0. The kf_0 are all distinct and form a sequence of discrete numbers. As $T \to \infty$, f_0 becomes smaller and smaller so that the kf_0 become almost continuous. Thus, instead of finding X_k for each distinct k, we would like to find $X(f)$ for each frequency f, where f is continuous and can assume any value now. What we would like to do is to generalize the formula for the Fourier series coefficients in (3-90). This generalization is accomplished by making the integral as a function of frequency. In view of Equation (3-90), we define the function of frequency as

$$X(f) = \int_{-\infty}^{\infty} x(t)e^{-j2\pi ft}dt. \tag{3-91}$$

We say that $X(f)$ is the Fourier transform of $x(t)$ which transforms $x(t)$ from the time domain to the frequency domain. In the following, we shall present the inverse Fourier transform which transforms $X(f)$ back to $x(t)$.

Using Equation (3-91), we can rewrite Equation (3-90) as

$$X_k = \frac{1}{T}X(kf_0). \tag{3-92}$$

Equation (3-92) shows that the function $X(f)$ is the envelope of the Fourier series coefficients X_k scaled by T. Substituting Equation (3-92) and $f_0 = 1/T$ into Equation (3-88), we get

$$\bar{x}(t) = \sum_{k=-\infty}^{\infty} \frac{1}{T}X(kf_0)e^{j2\pi kf_0t} = \sum_{k=-\infty}^{\infty} X(kf_0)e^{j2\pi kf_0t}f_0. \tag{3-93}$$

Taking the limit in Equation (3-93) and using Equation (3-87), we obtain

$$x(t) = \lim_{T\to\infty} \bar{x}(t) = \lim_{T\to\infty} \sum_{k=-\infty}^{\infty} X(kf_0)e^{j2\pi kf_0t}f_0. \tag{3-94}$$

As we discussed before, as $T \to \infty, f_0 \to 0$. That is, in the limit, we can conclude that $kf_0 \to f, f_0 \to df$, and the summation in Equation (3-94) becomes an integral. From Equation (3-94), we have

$$x(t) = \int_{-\infty}^{\infty} X(f)e^{j2\pi ft}df. \tag{3-95}$$

We say that $x(t)$ is the inverse Fourier transform of $X(f)$ which transforms $X(f)$ from the frequency domain back to the time domain.

Hence, Equations (3-91) and (3-95) represent the generalization of Fourier series to nonperiodic signals. This is called *Fourier transform*. We summarize these observations with the following definition.

Definition III

Suppose that $x(t), -\infty < t < \infty$, is a signal such that it is absolutely integrable, that is:

$$\int_{-\infty}^{\infty} |x(t)| dt < \infty.$$

Then the *Fourier transform* of $x(t)$ is defined as

$$X(f) = \int_{-\infty}^{\infty} x(t) e^{-j2\pi ft} dt \triangleq F\{x(t)\}, \qquad (3\text{-}96)$$

and the *inverse Fourier transform* is defined as

$$x(t) = \int_{-\infty}^{\infty} X(f) e^{j2\pi ft} df \triangleq F^{-1}\{X(f)\}. \qquad (3\text{-}97)$$

The one-to-one relationship between a signal $x(t)$ and its Fourier transform $X(f)$ forms a Fourier transform pair with the notation

$$x(t) \underset{F^{-1}}{\overset{F}{\rightleftharpoons}} x(f). \qquad (3\text{-}98)$$

Let us examine Equation (3-96). This equation defines the Fourier transform. As indicated by Equation (3-96), the Fourier transform transforms a function $x(t)$, a function in time, to $X(f)$, a function in frequency. *Thus the Fourier transform transforms a function in the time domain into a function in the frequency domain. Conversely, the inverse Fourier transform transforms a function in the frequency domain into a function in the time domain, as indicated in Equation (3-97).*

Note that the variable f in Equations (3-96) and (3-97) is a continuous variable. That is, for every frequency f, we have a corresponding $X(f)$. This is different from the Fourier series case where frequencies are defined by discrete kf_0 where $f_0 = 1/T$ is the fundamental frequency.

Example 3-12

Consider the rectangular pulse depicted in Figure 3-17(a) which is defined as

$$x(t) = \begin{cases} 1, & -T_s \leq t \leq T_s \\ 0, & \text{otherwise.} \end{cases}$$

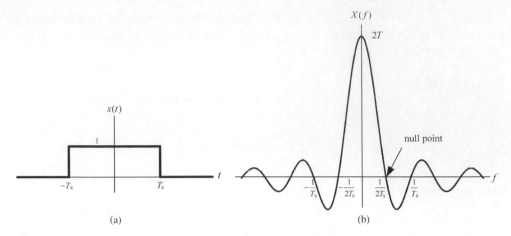

Figure 3-17   A rectangular pulse function $x(t)$: (a) $x(t)$ and (b) the Fourier transform of $x(t)$

We want to find the Fourier transform of $x(t)$. This rectangular pulse $x(t)$ is absolutely integrable provided $0 \leq T < \infty$. For $f \neq 0$, we have

$$
\begin{aligned}
X(f) &= \int_{-\infty}^{\infty} x(t)e^{-j2\pi ft}\,dt \\
&= \int_{-T_s}^{T_s} e^{-j2\pi ft}\,dt \\
&= -\frac{1}{j2\pi f}e^{-j2\pi ft}\big|_{-T_s}^{T_s} \\
&= \frac{1}{\pi f}\sin(2\pi fT_s).
\end{aligned}
\tag{3-99}
$$

For $f = 0$, the integral simplifies to become

$$
X(f) = \int_{-\infty}^{\infty} x(t)e^{-j2\pi ft}\,dt = \int_{-T}^{T} dt = 2T_s.
\tag{3-100}
$$

Using L'Hôpital's rule, we can see that

$$
\lim_{f \to 0}\frac{1}{\pi f}\sin(2\pi fT_s) = \lim_{f \to 0} 2T_s\cos(2\pi fT_s) = 2T_s.
\tag{3-101}
$$

Thus we write the Fourier transform of $x(t)$ for all f as

$$
X(f) = \frac{1}{\pi f}\sin(2\pi fT_s),
\tag{3-102}
$$

with the understanding that the value at $f = 0$ is obtained by evaluating a limit. In this case, $X(f)$ is real. It is depicted in Figure 3-17(b).

Consider Figure 3-17(b). Note that the Fourier transform spectrum changes as the width of the pulse changes. *If T_s is large, $1/2T_s$ is small, so only low-frequency components are significant. If T_s is very small, $1/2T_s$ is large, so there will be many high-frequency components which are still significant.* As indicated in Figure 3-17(b), the first point at which the Fourier transform function $X(f)$ intersects the frequency axis is called the *null point*. If the null point is large, this indicates that a large amount of frequencies are involved. On the other hand, if the null point is small, this means that only a narrow band of low frequencies is involved. Note that, in real practice, digital signals are transmitted in the form of pulses. *If we require the transmission system to be of high bit rate, the width of the pulse width will necessarily be very small – which implies the existence of many high-frequency signals which cannot be ignored. This will affect the transmission line performance.*

Example 3-13
Let us consider the function in Figure 3-17(a) again. Suppose we perform a time-shift of the function as shown in Figure 3-18. The Fourier transform of this function is as follows:

$$
\begin{aligned}
X(f) &= \int_{-\infty}^{\infty} x(t)e^{-j2\pi ft}dt \\
&= \int_{0}^{2T_s} e^{-j2\pi ft}dt \\
&= -\frac{1}{j2\pi f}e^{-j2\pi ft}\Big|_{0}^{2T_s} \\
&= \frac{(1 - e^{-j4\pi fT_s})}{j2\pi f} \\
&= \frac{e^{-j2\pi fT_s}}{j2\pi f}\left(e^{j2\pi fT_s} - e^{-j2\pi fT_s}\right) \\
&= e^{-j2\pi fT_s}\frac{1}{\pi f}\sin(2\pi fT_s)
\end{aligned}
\tag{3-103}
$$

Comparing Equation (3-103) with Equation (3-99), one can see that the Fourier transform of this function is almost the same as the original function without the time shift, except that there is a multiplier term $e^{-j2\pi fT_s}$ to the original Fourier transform. This phenomenon will be explained later in the next section.

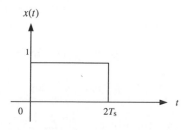

Figure 3-18   A function obtained by performing a time shift on the function in Figure 3-17

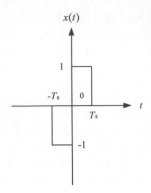

Figure 3-19   An odd function

Example 3-14

Consider the function in Figure 3-19. The Fourier transform of the function is computed as follows:

$$
\begin{aligned}
X(f) &= \int_{-\infty}^{\infty} x(t)e^{-j2\pi ft}\,dt \\
&= \int_{-T_s}^{0} -e^{-j2\pi ft}\,dt + \int_{0}^{T_s} e^{-j2\pi ft}\,dt \\
&= \frac{e^{-j2\pi ft}}{j2\pi f}\Big|_{-T_s}^{0} - \frac{e^{-j2\pi ft}}{j2\pi f}\Big|_{0}^{T_s} \\
&= \frac{2 - (e^{-j2\pi fT_s} + e^{j2\pi fT_s})}{j2\pi f} \\
&= \frac{1 - \cos(2\pi fT_s)}{j\pi f} \\
&= 2\frac{\sin^2(\pi fT_s)}{j\pi f}.
\end{aligned}
\tag{3-104}
$$

3.5 Physical Meaning of Fourier Transform

Equation (3-96) indicates that the Fourier transform transforms a function $x(t)$ in the time domain into a function $X(f)$ in the frequency domain. That is, for every frequency f, there is a corresponding frequency response $X(f)$. *Also, for every frequency f, there is also a negative frequency* $-f$. The meaning of $X(f)$ can be seen by the following equation:

$$
X(-f)e^{-j2\pi ft} + X(f)e^{j2\pi ft}.
\tag{3-105}
$$

Let us consider Equation (3-102). In this case, we can easily see that

$$
X(-f) = X(f) = \frac{\sin(2\pi fT_s)}{\pi f}.
$$

Thus:

$$
\begin{aligned}
X(-f)e^{-j2\pi ft} &+ X(f)e^{j2\pi ft} \\
&= X(f)(e^{-j2\pi ft} + e^{j2\pi ft}) \\
&= \frac{2}{\pi f}\sin(2\pi fT_s)\cos(2\pi ft)
\end{aligned}
\tag{3-106}
$$

Equation (3-106) is very similar to Equation (3-68), which should not be surprising. Consider Equation (3-104). In this case, we can show that $X(-f) = -X(f)$ and

$$
\begin{aligned}
X(-f)e^{-j2\pi ft} &+ X(f)e^{j2\pi ft} \\
&= X(f)(e^{j2\pi ft} - e^{-j2\pi ft}) \\
&= \frac{2\sin^2(\pi fT_s)}{j\pi f}2j\sin(2\pi ft) \\
&= \frac{4\sin^2(\pi fT_s)}{\pi f}\sin(2\pi ft).
\end{aligned}
\tag{3-107}
$$

Equation (3-107) is exactly the same as a general term in Equation (3-75) as expected. Finally, let us consider Equation (3-103). In this case

$$
\begin{aligned}
X(-f)e^{-j2\pi ft} &+ X(f)e^{j2\pi ft} \\
&- e^{j2\pi fT_s}\frac{\sin(2\pi fT_s)}{\pi f}e^{-j2\pi ft} \mid e^{-j2\pi fT_s}\frac{\sin(2\pi fT_s)}{\pi f}e^{j2\pi ft} \\
&= \frac{\sin(2\pi fT_s)}{\pi f}(e^{j2\pi f(t-T_s)} + e^{-j2\pi f(t-T_s)}) \\
&= \frac{\sin(2\pi fT_s)2\,\cos(2\pi f(t-T_s))}{\pi f} \\
&= \frac{2}{\pi f}\sin(2\pi fT_s)\cos(2\pi f(t-T_s))
\end{aligned}
\tag{3-108}
$$

It is interesting to compare Equations (3-108) and (3-106). One can see that Equation (3-107) can be obtained simply by substituting $t = t - T_s$ into Equation (3-106). This is expected because the function related to (3-108) is a time shift of that related to (3-106). The reader is encouraged to consult Figures 3-17(a) and 3-18.

By examining Equation (3-97), we note that every $X(f)$ is associated with $e^{j2\pi ft}$. It is easy to prove that, if $X(f) = a + jb = re^{j\theta}$, $X(-f) = a - jb = re^{-j\theta}$ where $\theta = \tan^{-1}(b/a)$. Thus, for every f, as we did before, we can easily prove that

$$
\begin{aligned}
X(-f)e^{-j2\pi ft} &+ X(f)e^{j2\pi ft} \\
&= re^{-j\theta}e^{-j2\pi ft} + re^{j\theta}e^{j2\pi ft} \\
&= re^{-j(2\pi ft+\theta)} + re^{j(2\pi ft+\theta)} \\
&= 2r\,\cos(2\pi ft + \theta) \\
&= 2|X(f)|\,\cos(2\pi ft + \theta)
\end{aligned}
\tag{3-109}
$$

where $\theta = \tan^{-1}(b/a)$.

Essentially, *the Fourier transform produces a function $X(f)$ in the frequency domain from a function $x(t)$ in the time domain. Both variables t and f are continuous. The function $x(t)$ consists of an infinite series of $X(f)$ and, for every f, $|X(f)|$ indicates the strength, or intensity, of a corresponding $\cos(2\pi ft + \theta)$ function and $\theta = \tan^{-1}(b/a)$ indicates a phase shift. The most significant point is that each frequency f is associated with a cosine function.*

We may conclude this section by saying that *the Fourier transform can be viewed as an analysis tool such that it produces its frequency component for every frequency f. Although $X(f)$ can be complex, it still corresponds to a cosine function with a certain phase shift.*

3.6 Properties of the Fourier Transform

Several useful properties of the Fourier transform will be presented here. These properties provide us with a significant amount of insight into the transform and into the relationship between the time-domain and frequency-domain descriptions of a signal.

1. Linearity
Consider $z(t) = ax(t) + by(t)$ as a linear combination of two signals, $x(t)$ and $y(t)$, where a and b are scalars. The Fourier transform of $z(t)$ is computed in terms of the Fourier transforms of $x(t)$ and $y(t)$ as follows:

$$
\begin{aligned}
Z(f) &= F\{ax(t) + by(t)\} \\
&= aF\{x(t)\} + bF\{y(t)\} \\
&= aX(f) + bY(f).
\end{aligned}
\tag{3-110}
$$

The linear property can be easily extended to a linear combination of any arbitrary number of signals. This property is obviously valid for the inverse Fourier transform.

2. Time Shift
Let $z(t) = x(t - t_0)$ be a time-shifted version of $x(t)$, where t_0 is a real number. The goal is to relate the Fourier transform of $z(t)$ to the Fourier transform of $x(t)$. We have

$$
Z(f) = \int_{-\infty}^{\infty} z(t)e^{-j2\pi ft}dt = \int_{-\infty}^{\infty} x(t - t_0)e^{-j2\pi ft}dt.
\tag{3-111}
$$

Performing the change of variable $\tau = t - t_0$, we obtain

$$
\begin{aligned}
Z(f) &= \int_{-\infty}^{\infty} x(\tau)e^{-j2\pi f(\tau + t_0)}d\tau \\
&= e^{-j2\pi ft_0} \int_{-\infty}^{\infty} x(\tau)e^{-j2\pi f\tau}d\tau \\
&= e^{-j2\pi ft_0} X(f).
\end{aligned}
\tag{3-112}
$$

The result of time-shifting by t_0 is to multiply the transform by $e^{-j2\pi f t_0}$. This implies that a signal which is shifted in time does not change the magnitude of its Fourier transform, but induces in its transform a phase shift, namely $-2\pi f t_0$.

A useful related formula is as follows:

$$F^{-1}\{e^{-j2\pi f t_0}X(f)\} = x(t - t_0).\tag{3-113}$$

The reader may be puzzled by what $e^{-j2\pi f t_0}X(f)$ means. Let $X'(f) = e^{-j2\pi f t_0}X(f)$ and $X(f) = re^{j\theta}$, where r and θ are functions of f. Then $X(-f) = re^{-j\theta}$ because $X(-f)$ and $X(f)$ are conjugate to each other, $X'(f) = re^{-j2\pi f t_0}e^{j\theta}$ and $X'(-f) = re^{j2\pi f t_0}e^{-j\theta}$. We now compute

$$
\begin{aligned}
e^{j2\pi ft}&X'(f) + e^{-j2\pi ft}X'(-f)\\
&= e^{j2\pi f(t-t_0)}X(f) + e^{-j2\pi f(t-t_0)}X(-f)\\
&= r(e^{j(2\pi f(t-t_0)+\theta)} + e^{-j(2\pi f(t-t_0)+\theta)})\\
&= 2r\,\cos(2\pi f(t - t_0) + \theta)
\end{aligned}
\tag{3-114}
$$

Equation (3-114) shows that the multiplication of $X(f)$ by $e^{-j2\pi f t_0}$ results in time shift in the time domain. The reader is encouraged to compare Equations (3-103) and (3-102). One can actually obtain (3-103) directly by applying (3-112) to (3-102).

3. Frequency Shift

In Property 2 above, we considered the effect of a time shift on the frequency-domain representation. Now we consider the effect of a frequency shift on the time-domain signal. Given $x(t)$, its Fourier transform is $X(f)$. Suppose we perform a frequency shift. That is, we have $X(f - \alpha)$. Our question is: What is its corresponding inverse transform, namely in the time domain, of $X(f - \alpha)$? That is, we would like to express the inverse Fourier transform of $Z(f) = X(f - \alpha)$ in terms of $x(t)$, where α is a real number. By the Fourier transform definition, we have

$$z(t) = \int_{-\infty}^{\infty} Z(f)e^{j2\pi ft}\,df = \int_{-\infty}^{\infty} X(f - \alpha)e^{j2\pi ft}\,df.$$

Performing the substitution of variable $v = f - \alpha$, we obtain

$$
\begin{aligned}
z(t) &= \int_{-\infty}^{\infty} X(v)e^{j2\pi(v+\alpha)t}\,dv\\
&= e^{j2\pi\alpha t}\int_{-\infty}^{\infty} X(v)e^{j2\pi vt}\,dv\\
&= e^{j2\pi\alpha t}x(t).
\end{aligned}
\tag{3-115}
$$

Hence the result of frequency-shifting by α corresponds to multiplication in the time domain by a complex sinusoid $e^{j2\pi\alpha t}$.

A useful related formula is as follows:

$$F\{e^{j2\pi\alpha t}x(t)\} = X(f - \alpha).\tag{3-116}$$

The meaning of Equation (3-115) might be puzzling because it is hard to comprehend the physical meaning of $e^{j2\pi\alpha t}x(t)$. Note that $x(t)$ is a real function. The function $e^{j2\pi\alpha t}x(t)$ contains an imaginary part. What does this imaginary part mean? This can be explained by noting

$$e^{j2\pi\alpha t} + e^{-j2\pi\alpha t} = 2\cos(2\pi\alpha t)$$

and

$$(e^{j2\pi\alpha t} + e^{-j2\pi\alpha t})x(t) = 2\cos(2\pi\alpha t)x(t).$$

Therefore:

$$
\begin{aligned}
F(\cos(2\pi\alpha t)x(t)) \\
= \frac{1}{2}(F\{e^{j2\pi\alpha t}x(t) + e^{-j2\pi\alpha t}x(t)\}) \\
= \frac{1}{2}(X(f - \alpha) + X(f + \alpha))
\end{aligned}
\tag{3-117}
$$

Equation (3-117) indicates that if we multiply a function $x(t)$ by $\cos(2\pi\alpha t)$, the net effect is that the frequencies of the original signal in $x(t)$ will be shifted. For each frequency f in $x(t)$, there will be two new frequencies: $f - \alpha$ and $f + \alpha$. We will elaborate this when we introduce amplitude modulation later.

If the reader is still puzzled by the above discussion, we may consider the problem in another way. The Fourier transform informs us that a function $x(t)$ consists of cosine functions. Consider an arbitrary function $x(t) = \cos(2\pi ft + \theta)$. If $x(t)$ is multiplied by $\cos(2\pi\alpha t)$, $\cos(2\pi ft + \theta)$ becomes $\cos(2\pi ft + \theta)\cos(2\pi\alpha t)$. But, by consulting equations in Appendix A, we may state:

$$
\begin{aligned}
\cos(2\pi ft + \theta)\cos(2\pi\alpha t) \\
= \frac{1}{2}\{\cos(2\pi(f + \alpha)t + \theta) + \cos(2\pi(f - \alpha)t + \theta)\}.
\end{aligned}
\tag{3-118}
$$

Using this line of reasoning we can see that each frequency f becomes $f + \alpha$ and $f - \alpha$.

Thus, although $e^{j2\pi\alpha t}x(t)$ is not physically meaningful, $\cos(2\pi\alpha t)x(t)$ has physical meaning. As we shall see in the next chapter, when we modulate a signal $x(t)$, we multiply it by $\cos(2\pi f_c t)$ for some frequency f_c. Therefore Equation (3-116) is useful for us to find the Fourier transform of $\cos(2\pi f_c t)x(t)$.

4. Convolution

The convolution $z(t)$ of two signals $x(t)$ and $y(t)$ is defined as follows:

$$z(t) \overset{\Delta}{=} x(t)^*y(t) = \int_{-\infty}^{\infty} x(\tau)y(t - \tau)d\tau.
\tag{3-119}$$

The convolution operation is commutative, namely, $x(t)^*y(t) = y(t)^*x(t)$. In the following, let us try to find the Fourier transform of $z(t)$.

$$
\begin{aligned}
Z(f) &= \int_{\infty}^{\infty} \left[\int_{-\infty}^{\infty} x(\tau)y(t-\tau)d\tau \right] e^{-j2\pi ft} dt \\
&= \int_{-\infty}^{\infty} x(\tau) \left[\int_{-\infty}^{\infty} y(t-\tau)e^{-j2\pi ft} dt \right] d\tau \\
&= \int_{-\infty}^{\infty} x(\tau)Y(f)e^{-j2\pi\tau} d\tau \qquad\qquad (3\text{-}120) \\
&= \left[\int_{-\infty}^{\infty} x(\tau)e^{-j2\pi\tau} d\tau \right] Y(f) \\
&= X(f)Y(f).
\end{aligned}
$$

Thus we have

$$
z(t) = x(t) * y(t) \underset{F^{-1}}{\overset{F}{\rightleftharpoons}} Z(f) = X(f)Y(f). \qquad (3\text{-}121)
$$

This is an important property which can be used to analyze the input–output behavior of a linear system in the frequency domain using multiplication of Fourier transforms instead of convolving time signals. The physical meaning of convolution will become clear after we have introduced the concept of modulation.

5. Modulation

The modulation property here refers to the multiplication of two signals; one of the signals changes or "modulates" the amplitude of the other. We want to express the Fourier transform of the product $z(t) = x(t)y(t)$ in terms of the Fourier transforms of the signals $x(t)$ and $y(t)$. Let us represent $x(t)$ and $y(t)$ in terms of their Fourier transforms respectively in the following:

$$
x(t) = \int_{-\infty}^{\infty} X(v)e^{j2\pi vt} dv
$$

$$
y(t) = \int_{-\infty}^{\infty} Y(\eta)e^{j2\pi\eta t} d\eta.
$$

The product term, $z(t)$, may thus be written in the form

$$
\begin{aligned}
z(t) &= \left(\int_{-\infty}^{\infty} X(v)e^{j2\pi vt} dv \right) \left(\int_{-\infty}^{\infty} Y(\eta)e^{j2\pi\eta t} d\eta \right) \\
&= \int_{-\infty}^{\infty} \int_{-\infty}^{\infty} X(v)Y(\eta)e^{j2\pi(v+\eta)t} dv d\eta. \qquad (3\text{-}122)
\end{aligned}
$$

Performing a change of variable on η and substituting $\eta = f - v$, we obtain

$$
z(t) = \int_{-\infty}^{\infty} \left(\int_{-\infty}^{\infty} X(v)Y(f-v))dv \right) e^{j2\pi ft} df.
$$

The inner integral over v in the above equation represents the convolution of $X(f)$ and $Y(f)$. Therefore, the modulated signal $z(t)$ can be rewritten as

$$z(t) = \int_{-\infty}^{\infty} (X(f) * Y(f)) e^{j2\pi ft} df, \tag{3-123}$$

which tells us that $z(t)$ is the inverse Fourier transform of $X(f) * Y(f)$. To put it another way, we may say that the Fourier transform of $z(t) = x(t)y(t)$ is $X(f) * Y(f)$. This result shows that the multiplication of two signals in the time domain leads to the convolution of their transforms in the frequency domain, as shown by

$$z(t) = x(t)y(t) \underset{F^{-1}}{\overset{F}{\rightleftharpoons}} Z(f) = X(f) * Y(f). \tag{3-124}$$

The last two properties are very important in Fourier transform analysis. We can conclude that convolution in the time domain is transformed to modulation in the frequency domain, and modulation in the time domain is transformed to convolution in the frequency domain, which are shown in Equations (3-121) and (3-124) respectively.

Since it is by no means easy to understand the physical meaning of convolution, we now give an informal explanation of it through an example.

Example 3-15

Let $x(u) = 1$ for $-1 \le u \le 1$ and $x(u) = 0$ every where else, and let $y(u) = 1 - u$ for $0 \le u \le 1$ and $y(u) = 0$ every where else. Functions $x(u)$ and $y(u)$ are illustrated in Figures 3-20(a) and 3-20(b) respectively.

Figure 3-20(c) shows $y(t - u) = 1 - (t - u) = (1 - t) + u$ for some constant t. As shown, $y(t - u)$ is obtained by two operations. First, $y(u)$ is reversed. Second, $y(u)$ is slid either to the left (when t is negative) or to the right (when t is positive). We may therefore imagine that $y(t - u)$ is slid from minus infinity to plus infinity as we compute

$$z(t) = x(t) * y(t) = \int_{-\infty}^{\infty} x(u)y(t - u)du.$$

There are the following different cases.

Case 1: $-\infty \le u < -1(-\infty \le t < -1)$. In this case $z(t) = 0$.

Case 2: $-1 \le u < 0(-1 \le t < 0)$. In this case, $x(u) = 1$ intersects with $y(t - u)$ and

$$z(t) = x(t) * y(t) = \int_{-\infty}^{\infty} x(u)y(t - u)du$$

starts to have some value as this actually computes the area corresponding to the intersection of $x(u)$ and $y(t - u)$ for a particular t, as illustrated in Figure 3-20(d).

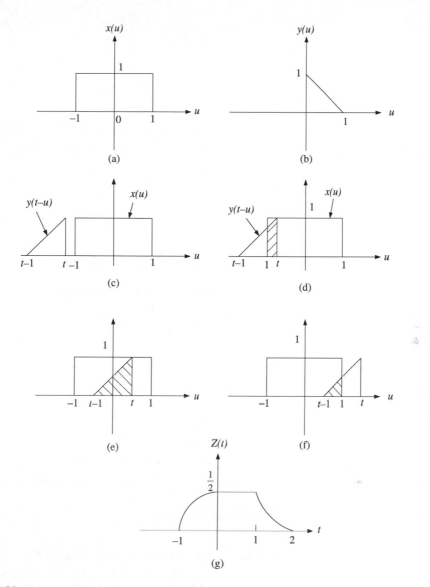

Figure 3-20 An example of convolution: (a) $x(u)$; (b) $y(u)$; (c) Case 1: $y(t-u)$; (d) Case 2; (e) Case 3; (f) Case 4; (g) $z(t) = x(t)^*y(t)$

Thus, in this region, we integrate from $u = -1$ to $u = t$ and we have

$$z(t) = x(t) * y(t) = \int_{-1}^{t} 1(1 - (t - u))du$$

$$= (1 - t)u + \frac{1}{2}u^2 \Big|_{-1}^{t}$$

$$= (1-t)t + \frac{1}{2}t^2 - (1-t)(-1) - \frac{1}{2}(-1)^2$$

$$= t - t^2 + \frac{1}{2}t^2 + 1 - t - \frac{1}{2}$$

$$= \frac{1}{2}(1 - t^2).$$

Case 3: $0 \le u \le 1 (0 \le t \le 1)$. As $y(t-u)$ slides from $t = 0$ to $t = 1$, the area intersected by $x(u)$ and $y(t-u)$ remains the same, as illustrated in Figure 3-20(e). At $t = 0$, $z(t) = x(t) * y(t) = 1/2$. Thus $z(t)$ is 1/2 in this region.

Case 4: $1 \le u \le 2 (1 \le t \le 2)$. In this case, as illustrated in Figure 3-20(f), we integrate from $u = t - 1$ to $u = 1$.

$$z(t) = \int_{t-1}^{1} x(u)y(t-u)du$$

$$= \int_{t-1}^{1} (1 - t + u)du$$

$$= u - tu + \frac{1}{2}u^2 \Big|_{t-1}^{1}$$

$$= \left(1 - t + \frac{1}{2}\right) - \left((t-1) - t(t-1) + \frac{1}{2}(t-1)^2\right)$$

$$= \frac{1}{2}t^2 - 2t + 2$$

$$= \frac{1}{2}(t-2)^2.$$

Case 5: $2 \le u \le \infty (2 \le t \le \infty)$. In this case $z(t) = 0$.

The final $z(t) = x(t) * y(t)$ is now illustrated in Figure 3-20(g). Note that the convolution has some smoothing effect. $x(u)$ originally has sharp edges. Now they are gone. Also, the function is flattened and widened. Originally, the width of $x(u)$ was 2; it is now 3. The height of $x(u)$ is also reduced from 1 to 1/2. These properties are significant, as will become clear when we introduce the spread spectrum techniques.

As we indicated before, convolution is symmetric in the sense that

$$x(t) * y(t) = y(t) * x(t).$$

Therefore, for the case in Example 3-15, we may also fix $y(u)$ and slide $x(u)$ from $-\infty$ to the right. There are again three cases for $x(t - u)y(u)$ not being 0, as illustrated in Figures 3-21(a), 3-21(b) and 3-21(c). The reader is encouraged to prove that the result will be the same as obtained in Example 3-15.

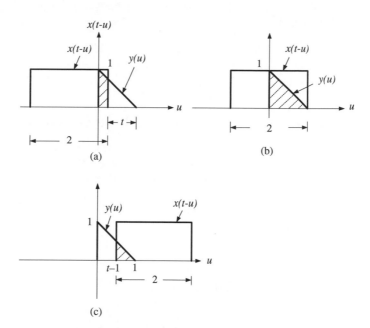

Figure 3-21   Convolution from another perspective

In convolution $x(t) * y(t)$, we may consider $x(u)$ as a scanner scanning the function $y(u)$ from left to right and reporting the result. This concept will be useful when we study amplitude modulation.

Let us now summarize the properties mentioned in this section as follows:

1. Linearity

$$F(ax(t) + by(t)) = aX(f) + bY(f). \tag{3-125}$$

$$F^{-1}(aX(f) + b\,(Y(f)) = ax(t) + by(t). \tag{3-126}$$

2. Time Shift

$$F(x(t - t_0)) = e^{-j2\pi f t_0}X(f). \tag{3-127}$$

$$F^{-1}(e^{-j2\pi f t_0}X(f)) = x(t - t_0). \tag{3-128}$$

3. Frequency Shift

$$F^{-1}(X(f - \alpha)) = e^{j2\pi\alpha t}x(t). \tag{3-129}$$

$$F(e^{j2\pi\alpha t}x(t)) = X(f - \alpha). \tag{3-130}$$

4. Convolution

$$F(x(t) * y(t)) = X(f)Y(f). \tag{3-131}$$

$$F^{-1}(X(f)Y(f)) = x(t) * y(t). \tag{3-132}$$

5. Modulation

$$F(x(t)y(t)) = X(f) * Y(f). \tag{3-133}$$

$$F^{-1}(X(f) * Y(f)) = x(t)y(t). \tag{3-134}$$

3.7 Fourier Transform Representations for Periodic Signals

Recall that the Fourier series has been derived as the Fourier representation for periodic signals. Strictly speaking, the Fourier transform does not converge for periodic signals because they are usually not absolutely integrable. Nonetheless, it would be useful to have a "Fourier transform" for these signals. By incorporating *impulses* into the Fourier transform in the appropriate manner, we may develop Fourier transform representations for periodic signals. Hence, we introduce the ***impulse function*** first.

The unit impulse function, commonly denoted by $\delta(t)$, is defined by the following pair of relations:

$$\delta(t) = 0 \text{ for } t \neq 0 \tag{3-135}$$

$$\int_{-\infty}^{\infty} \delta(t)dt = 1. \tag{3-136}$$

Equation (3-135) says that the impulse $\delta(t)$ is zero everywhere except at the origin. Equation (3-136) says that the total area under the unit impulse function is unity. The impulse $\delta(t)$ is also referred to as the ***Dirac delta function***.

An important property of the unit impulse is the ***sifting property***. This property incorporates Equations (3-135) and (3-136) into a single relation, described as

$$\int_{-\infty}^{\infty} x(t)\delta(t - t_0)dt = x(t_0). \tag{3-137}$$

The operation indicated on the left-hand-side of (3-137) sifts the value $x(t_0)$ of the function $x(t)$ at time $t = t_0$ by the delta function $\delta(t - t_0)$.

A graphical description to visualize $\delta(t)$ shows it as the limiting form of a rectangular pulse of unit area, as illustrated in Figure 3-22. The smaller the duration becomes, the better the rectangular pulse approximates the impulse. Specifically, the duration of the pulse is decreased and its amplitude is increased such that the area under the pulse is maintained constant at unity. The area under the pulse defines the strength of the impulse. Therefore, when we speak of the impulse function $\delta(t)$, in effect we are saying that its strength is unity.

The graphical symbol for an impulse is like an arrow depicted in Figure 3-22. The strength of the impulse is denoted by the label next to the arrow.

Example 3-16

In this example, we are interested in determining the Fourier transform of the unit impulse signal $x(t) = \delta(t)$. Using the sifting property of the unit impulse in Equation (3-137), we obtain

$$X(f) = \int_{-\infty}^{\infty} \delta(t)e^{-j2\pi ft}dt = e^{-j2\pi f0} = 1,$$

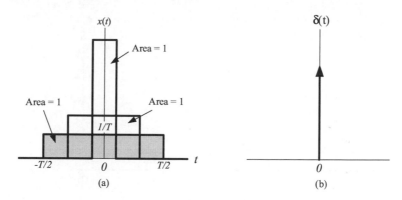

Figure 3-22   The Delta function: (a) evolution of a rectangular pulse of a unit area into an impulse of unit strength; (b) graphical symbol for an impulse

which implies the following Fourier transform pair:

$$\delta(t) \underset{F^{-1}}{\overset{F}{\rightleftharpoons}} 1. \tag{3-138}$$

We are also interested in finding the inverse Fourier transform of $X(f) = \delta(f)$ (the unit impulse function in the frequency domain). It is computed as

$$x(t) = \int_{-\infty}^{\infty} \delta(f) e^{j2\pi ft} df = e^{j0t} = 1, \tag{3-139}$$

which yields the following Fourier transform pair:

$$1 \underset{F^{-1}}{\overset{F}{\rightleftharpoons}} \delta(f). \tag{3-140}$$

Now, let us explain what is meant by $F(1) = \delta(f)$ and $F(\delta(t)) = 1$. Interestingly, this can be best explained by considering the pulse function in Figure 3-17(a). Consider the following two cases.

Case 1: $T \to \infty$. In this case, the function becomes a constant as shown in Figure 3-23(a). In electrical engineering terms, this is a direct current (DC) whose frequency is 0. In this case, $x(t) = 1$. The Fourier transform spectrum of this function will become very narrow as the null point $1/2T$, illustrated in Figure 3-17(b), becomes extremely small, as shown in Figure 3-23(b). It thus can be easily seen that the Fourier transform of $x(t) = 1$ is $\delta(f)$. This explains why Equation (3-140) is valid.

Case 2: $T \to 0$. In this case, the function becomes an extremely narrow pulse as shown in Figure 3-24(a). Thus, we may say that $x(t) = \delta(t)$. For the corresponding Fourier transform, the null point $1/2T$ will become infinity, which means that the Fourier transform will become a constant as shown in Figure 3-24(b). This explains Equation (3-138).

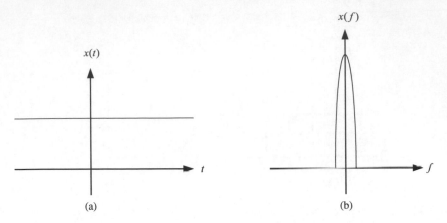

(a)   (b)

Figure 3-23 The Fourier transform of $x(t) = 1$: (a) the function; (b) its corresponding Fourier transform

Example 3-17

The Fourier transform of the signal $x(t) = e^{j2\pi\alpha t}$, where α is a real number, can be computed by

$$X(f) = F\{e^{j2\pi\alpha t}\}$$
$$= F\{e^{j2\pi\alpha t} \cdot 1\}. \tag{3-141}$$

According to Equation (3-116), the Fourier transform of $y(t)e^{j2\pi\alpha t}$ is equal to $Y(f - \alpha)$. In this case, $y(t) = 1$. According to Equation (3-140), $Y(f) = \delta(f)$ and $Y(f - \alpha) = \delta(f - \alpha)$. Thus, we have

$$F(e^{j2\pi\alpha t}) = \delta(f - \alpha). \tag{3-142}$$

Now we are going to derive the Fourier transform representations for periodic signals. The Fourier series representation for a periodic signal $x(t)$ defined in Definition II is

$$x(t) = \sum_{k=-\infty}^{\infty} X_k e^{j2\pi k f_0 t}, \tag{3-143}$$

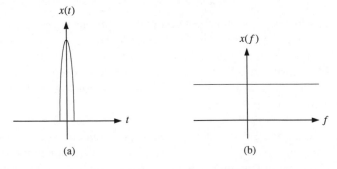

(a)   (b)

Figure 3-24 The Fourier transform of $x(t) = \delta(t)$: (a) the function; (b) its corresponding Fourier transform

where f_0 is the fundamental frequency of $x(t)$. From Equation (3-110) and (3-142), the Fourier transform of this periodic signal $x(t)$ represented by its Fourier series can be computed as

$$X(f) = F\left\{\sum_{k=-\infty}^{\infty} X_k e^{j2\pi k f_0 t}\right\}$$

$$= \sum_{k=-\infty}^{\infty} X_k F\{e^{j2\pi k f_0 t}\}$$

$$= \sum_{k=-\infty}^{\infty} X_k \delta(f - k f_0). \qquad (3\text{-}144)$$

Hence the Fourier transform of a periodic signal is a series of impulses spaced by the fundamental frequency f_0. The kth impulse has strength X_k, where X_k is the kth Fourier series coefficient. Figure 3-25 illustrates this relationship. Since we have denoted the strength of impulses in the figures by their height as indicated by the labels on the vertical axis, we see that the shape of $X(f)$ is identical to that of X_k. Equation (3-144) also indicates how to convert Fourier transform to Fourier series representations for periodic signals, and vice versa. Given a Fourier transform consisting of impulses that are uniformly spaced on the f axis, we can also obtain Fourier series coefficients. The fundamental frequency corresponds to the spacing between impulses.

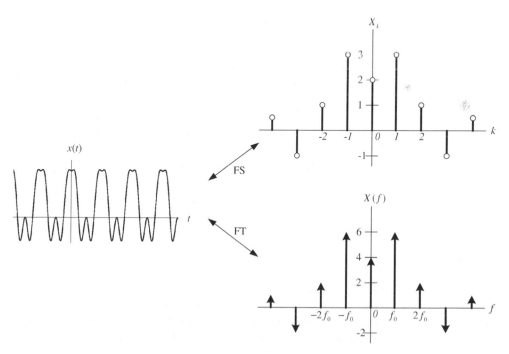

Figure 3-25   The Fourier series and Fourier transform representations for a periodic signal

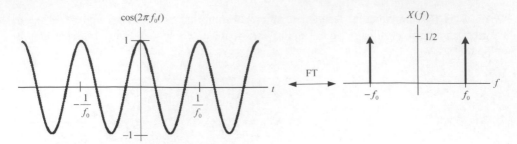

Figure 3-26   The Fourier transform of a cosine function

Example 3-18
From Equations (3-57) and (3-142), the Fourier transform of the periodic signal
$x(t) = \cos 2\pi f_0 t$ with frequency f_0 can be computed by

$$
\begin{aligned}
X(f) &= F\{\cos(2\pi f_0 t)\} \\
&= F\left\{\frac{1}{2}(e^{j2\pi f_0 t} + e^{-j2\pi f_0 t})\right\} \\
&= \frac{1}{2}F\{e^{j2\pi f_0 t}\} + \frac{1}{2}F\{e^{-j2\pi f_0 t}\} \\
&= \frac{1}{2}(\delta(f - f_0) + \delta(f + f_0)).
\end{aligned}
\tag{3-145}
$$

The result is illustrated in Figure 3-26. Similarly, using Equations (3-57) and (3-141), the
Fourier transform of the periodic signal $x(t) = \sin(2\pi f_0 t)$ is calculated by

$$
\begin{aligned}
X(f) &= F\{\sin(2\pi f_0 t)\} \\
&= F\left\{\frac{1}{2j}(e^{j2\pi f_0 t} - e^{-j2\pi f_0 t})\right\} \\
&= \frac{1}{2j}F\{e^{j2\pi f_0 t}\} - \frac{1}{2j}F\{e^{-j2\pi f_0 t}\} \\
&= \frac{1}{2j}\delta(f - f_0) - \frac{1}{2j}\delta(f + f_0) \\
&= \frac{1}{2j}[\delta(f - f_0) - \delta(f + f_0)].
\end{aligned}
\tag{3-146}
$$

3.8 The Discrete Fourier Transform

We shall now introduce the concept of the discrete Fourier transform. In practice, we often face
the following situation: We are given a waveform $x(t)$, possibly representing a human voice.
Since we do not know the mathematical definition of $x(t)$, it is impossible for us to calculate the
Fourier transform $X(f)$. Thus we cannot perform a Fourier transform on this function.

Although we do not know the mathematical formula for $x(t)$, and therefore cannot calculate

$$X(f) = \int_{-\infty}^{\infty} x(t)e^{-j2\pi ft} dt,$$

we may still do the following. First we divide the entire time period of $x(t)$ into $n = 2^p$ divisions where p is an integer. Let us give an example. Suppose we are given a function $x(t) = \cos(2\pi t)$ and we are interested in the interval $0 \leq t < T$. Then we may sample the function $x(t)$ with sampling interval $T_s = T/n$. Instead of $x(t) = \cos(2\pi t)$, we shall use its corresponding sample function: $x(kT_s) = \cos(2\pi kT_s)$. Without risking ambiguity, we may denote this as $x(k) = x(kT_s)$ for simplicity. After having decided the number n, we may have n samples of $x(t)$. Let $k = 0, 1, 2, \cdots, n-1$ and let $a_k = x(k)$ for $k = 0, 1, 2, \cdots, n-1$. We may say that function $x(t)$ is now characterized by $a_0, a_1, \ldots, a_{n-1}$. It is easy to see intuitively the larger n is, the more accurately $x(t)$ is characterized.

Note that our goal is to compute

$$X(f) = \int_{-\infty}^{\infty} x(t)e^{-j2\pi ft} dt.$$

Since we cannot do that analytically, we are going to find an approximation to the solution. First, we have a fundamental frequency $f_0 = 1/T$. Since $X(f)$ is also a continuous function of f, we may also require to sample $X(f)$ in the frequency domain. We sample the frequency at if_0 for $i = 0, 1, 2, \ldots, n-1$. Let us suppose that we are interested in finding $X(if_0)$ and we have already obtained $a_0, a_1, \cdots, a_{n-1}$, where $a_k = x(k)$. Without ambiguity, we may denote $X(i) = X(if_0)$. Thus, we may represent $e^{-j2\pi ft}$ as $e^{-j2\pi if_0 \frac{kT}{n}} = e^{-j2\pi \frac{ikT}{nT}} = e^{-j2\pi \frac{ik}{n}}$. Given i, we can now perform the following:

$$x(0)e^{-j\frac{2\pi i(0)}{n}} + x(1)e^{-j\frac{2\pi i(1)}{n}} + \cdots + x(n-1)e^{-j\frac{2\pi i(n-1)}{n}}$$

$$= a_0 e^{-j\frac{2\pi i(0)}{n}} + a_1 e^{-j\frac{2\pi i(1)}{n}} + \cdots + a_{n-1} e^{-j\frac{2\pi i(n-1)}{n}}$$

$$= \sum_{k=0}^{n-1} a_k e^{-j\frac{2\pi ik}{n}}.$$

The above expression is close to $X(i)$ in the Fourier transform. We therefore define

$$A_i = X(i) = \sum_{k=0}^{n-1} a_k e^{-j\frac{2\pi ik}{n}}, \quad \text{for} \quad 0 \leq i \leq n-1. \tag{3-147}$$

Equation (3-147) is the **discrete Fourier transform of** $x(t)$. We are given the a_k for $k = 0, 1, 2, \cdots, n-1$, where $a_k = x(k)$, and our job is to find the A_i for $i = 0, 1, 2, \cdots, n-1$. Thus the discrete Fourier transform transforms a function in the time domain to a function in the frequency domain. The reader is encouraged to compare Equation (3-147) with (3-91). One can easily see the following:

1. There is a fundamental frequency f_0 which is equal to $f_0 = 1/T$.
2. A_i, for $i = 0, 1, \cdots, (n/2) - 1$, in Equation (3-147) is associated with frequency if_0. If the a_k are real numbers, it can be easily proved that A_0 and $A_{n/2}$ are real numbers, and $A_{n-i} = \overline{A_i}$ (A_{n-i} and A_i are conjugate) for $i = 1, 2, \cdots, n - 1$. Thus, if the a_k are real numbers, we may ignore the frequencies from $i = (n/2) + 1$ to $i = n - 1$. The reader will finally understand that this situation is similar to the case in the Fourier transform. In the Fourier transform, there are negative frequencies which do not have physical meaning. The only reason that negative frequencies exist in the Fourier transform is that they, together with positive frequencies, produce cosine functions. *In the discrete Fourier transform, if the a_k are real numbers, the frequencies from $i = (n/2) + 1$ to $i = n - 1$ do not have physical meaning either. They exist so that finally cosine functions can be produced*. This will be presented in the next section.
3. In general, A_i is a complex number. The magnitude of A_i, denoted as $|A_i|$, indicates the strength of a cosine function contained in $x(t)$ with frequency if_0.
4. The most significant property of the discrete Fourier transform is the property of the *inverse discrete Fourier transform* which is defined as follows. Suppose that we are given $A_0, A_1, \cdots, A_{n-1}$, then $a_k = x(k)$ can be found as follows:

$$a_k = \frac{1}{n} \sum_{i=0}^{n-1} A_i e^{j\frac{2\pi ik}{n}} \text{ for } 0 \leq k \leq n - 1. \tag{3-148}$$

Equation (3-148) indicates that, after we obtain the discrete Fourier transform coefficients A_i, we can use them to reconstruct the original function $x(k)$ through the inverse discrete Fourier transform. Thus the inverse Fourier transform transforms a function in the frequency domain into a function in the time domain. *The reader should note here that we cannot reconstruct the entire $x(t)$. We can only recover n points of $x(t)$, namely $a_k = x(k)$ for $k = 0, 1, 2, \cdots, n - 1$. Again, as we indicated before, if n is large, we can almost recover the entire $x(t)$.*
5. Note that $n = 2^p$, where p is an integer.

In the following, we shall show why Equation (3-148) is valid. We first substitute (3-147) into (3-148) and denote the result as b_i; that is:

$$b_i = \frac{1}{n} \sum_{k=0}^{n-1} A_k e^{j2\pi\frac{ki}{n}}$$

$$= \frac{1}{n} \sum_{k=0}^{n-1} \left(\sum_{i'=0}^{n-1} a_{i'} e^{-j2\pi\frac{ki'}{n}} \right) e^{j2\pi\frac{ki}{n}}$$

$$= \frac{1}{n} \sum_{i'=0}^{n-1} a_{i'} \left(\sum_{k=0}^{n-1} e^{-j2\pi\frac{(i-i')}{n}k} \right). \tag{3-149}$$

The term inside the brackets can be shown to be

$$\sum_{k=0}^{n-1} e^{-j2\pi\frac{(i-i')}{n}k} = \frac{1 - e^{-j2\pi(i-i')}}{1 - e^{-j2\pi(i-i')/n}}$$

$$= \begin{cases} n, & i - i' = 0, \pm n, \pm 2n, \dots \\ 0 & \text{otherwise.} \end{cases}$$ (3-150)

The first equation follows, since $\sum_{k=0}^{n-1} x^k = (1 - x^n)/(1 - x)$. Now since $0 \le i \le n - 1$ and $0 \le i' \le n - 1$, we have $-n + 1 \le i - i' \le n - 1$. Therefore for $0 \le i \le n - 1$ and $0 \le i' \le n - 1$, the above equation can be expressed as

$$\sum_{k=0}^{i-1} e^{-j2\pi\frac{(i-i')}{n}k} = \begin{cases} n, & i - i' = 0 \\ 0 & \text{otherwise} \end{cases}$$

$$= n\delta_{i-i'}.$$ (3-151)

Now, b_i can be expressed as

$$b_i = \frac{1}{n}\sum_{i'=0}^{n-1} a_{i'} n\delta(i - i').$$ (3-152)

Each term in the summation is zero, except the one with index $i' = i$. Therefore, it follows that

$$b_i = \frac{1}{n} a_i n = a_i.$$ (3-153)

Thus, we have proved that the inverse discrete Fourier transform works.

In the following, we shall use a simple case to illustrate the discrete Fourier transform. Consider the case where $n = 4$ and

$$e^{-j\frac{2\pi}{n}} = e^{-j\frac{2\pi}{4}} = e^{-j\frac{\pi}{2}} = \cos\frac{\pi}{2} - j\sin\frac{\pi}{2} = -j.$$

From Equation (3-147) we have:

$$A_0 = a_0 + a_1 + a_2 + a_3$$ (3-154)

$$A_1 = a_0 + a_1(-j) + a_2(-j)^2 + a_3(-j)^3$$
$$= a_0 - ja_1 - a_2 + ja_3$$ (3-155)

$$A_2 = a_0 + a_1(-j)^2 + a_2(-j)^4 + a_3(-j)^6$$
$$= a_0 - a_1 + a_2 - a_3$$ (3-156)

$$A_3 = a_0 + a_1(-j)^3 + a_2(-j)^6 + a_3(-j)^9$$
$$= a_0 + ja_1 - a_2 - ja_3.$$ (3-157)

Having found A_0, A_1, A_2 and A_3, we can inversely find a_0, a_1, a_2 and a_3 by Equation (3-148). Note that

$$e^{j\frac{2\pi}{n}} = e^{j\frac{\pi}{2}} = j.$$

For the coefficient a_0, we have

$$\frac{1}{4}(A_0 + A_1 + A_2 + A_3)$$

$$= \frac{1}{4}(a_0 + a_1 + a_2 + a_3 + a_0 - ja_1 - a_2 + ja_3 + a_0 - a_1 + a_2 - a_3 + a_0 + ja_1 - a_2 - ja_3)$$

$$= \frac{1}{4}(4a_0)$$

$$= a_0.$$

For the coefficient a_1, we have

$$\frac{1}{4}(A_0 + A_1(j) + A_2(j^2) + A_3(j^3))$$

$$= \frac{1}{4}(A_0 + jA_1 - A_2 - jA_3)$$

$$= \frac{1}{4}(a_0 + a_1 + a_2 + a_3 + ja_0 + a_1 - ja_2 - a_3 + (-a_0) + a_1 - a_2 + a_3 - ja_0 + a_1 + ja_2 + a_3)$$

$$= \frac{1}{4}(4a_1)$$

$$= a_1.$$

For the coefficient a_2, we have

$$\frac{1}{4}(A_0 + A_1(j^2) + A_2(j^4) + A_3(j^6))$$

$$= \frac{1}{4}(A_0 - A_1 + A_2 - A_3)$$

$$= \frac{1}{4}(a_0 + a_1 + a_2 + a_3 - a_0 + ja_1 + a_2 - ja_3 + a_0 - a_1 + a_2 - a_3 - a_0 - ja_1 + a_2 + ja_3)$$

$$= \frac{1}{4}(4a_2)$$

$$= a_2.$$

For the coefficient a_3, we have

$$\frac{1}{4}(A_0 + A_1(j^3) + A_2(j^6) + A_3(j^9))$$

$$= \frac{1}{4}(A_0 - jA_1 - A_2 + jA_3)$$

$$= \frac{1}{4}(a_0 + a_1 + a_2 + a_3 - ja_0 - a_1 + ja_2 + a_3 - a_0 + a_1 - a_2 + a_3 + ja_0 - a_1 - ja_2 + a_3)$$

$$= \frac{1}{4}(4a_3)$$

$$= a_3$$

We have shown that the inverse Fourier transform does recover the original a_k.

A straightforward approach to computing the discrete Fourier transform would require $O(n^2)$ steps. If the divide and conquer strategy is used, we can reduce the time-complexity to $O(n\log n)$. This technology is the so-called **fast Fourier transform**. The reader may ignore this if he/she deems it is appropriate to do so.

To simplify our discussion, let

$$w_n = e^{-j\frac{2\pi}{n}}.$$

Thus, from Eqation (3-147), the discrete Fourier transform is to calculate the following:

$$A_i = a_0 + a_1 w_n^i + a_2 w_n^{2i} + \cdots + a_{n-1} w_n^{(n-1)i}. \tag{3-158}$$

Let $n = 3$. Then, we have

$$A_0 = a_0 + a_1 + a_2 + a_3$$
$$A_1 = a_0 + a_1 w_4 + a_2 w_4^2 + a_3 w_4^3$$
$$A_2 = a_0 + a_1 w_4^2 + a_2 w_4^4 + a_3 w_4^6$$
$$A_3 = a_0 + a_1 w_4^3 + a_2 w_4^6 + a_3 w_4^9. \tag{3-159}$$

We may rearrange the above equations into the following forms:

$$A_0 = (a_0 + a_2) + (a_1 + a_3) \tag{3-160}$$
$$A_1 = (a_0 + a_2 w_4^2) + w_4(a_1 + a_3 w_4^2) \tag{3-161}$$
$$A_2 = (a_0 + a_2 w_4^4) + w_4^2(a_1 + a_3 w_4^4) \tag{3-162}$$
$$A_3 = (a_0 + a_2 w_4^6) + w_4^3(a_1 + a_3 w_4^6). \tag{3-163}$$

Since $w_n^2 = w_{n/2}$ and $w_n^{n+k} = w_n^k$, we have

$$A_0 = (a_0 + a_2) + (a_1 + a_3)$$
$$A_1 = (a_0 + a_2 w_2) + w_4(a_1 + a_3 w_2)$$
$$A_2 = (a_0 + a_2) + w_2(a_1 + a_3). \tag{3-164}$$
$$A_3 = (a_0 + a_2 w_4^2) + w_4^3(a_1 + a_3 w_4^2)$$
$$\quad = (a_0 + a_2 w_2) + w_4^3(a_1 + a_3 w_2)$$

Let

$$B_0 = a_0 + a_2$$
$$C_0 = a_1 + a_3$$
$$B_1 = a_0 + a_2 w_2$$
$$C_1 = a_1 + a_3 w_2.$$

$$(3\text{-}165)$$

Then:

$$A_0 = B_0 + w_4^0 C_0$$
$$A_1 = B_1 + w_4^1 C_1$$
$$A_2 = B_0 + w_4^2 C_0$$
$$A_3 = B_1 + w_4^3 C_1.$$

$$(3\text{-}166)$$

The above equation shows that the divide and conquer strategy can be elegantly applied to solve the discrete Fourier transform problem for $n = 3$. We merely have to calculate B_0, C_0, B_1 and C_1. The A_j can be easily obtained. In other words, once A_0 is obtained, A_2 can be obtained immediately. Similarly, once A_1 is obtained, A_3 can be obtained directly.

But, what are the B_i and C_i? Note that the B_i are the discrete Fourier transforms of the odd-numbered input data items, and the C_i are the discrete Fourier transform of the even-numbered input data items. This is the basis of applying the divide and conquer strategy to the discrete Fourier transform problem. We break a large problem into two subproblems, solve these subproblems recursively and then merge the solutions.

Now consider A_i in the general case:

$$\begin{aligned}
A_i &= a_0 + a_1 w_n^i + a_2 w_n^{2i} + \cdots + a_{n-1} w_n^{(n-1)i} \\
&= (a_0 + a_2 w_n^{2i} + a_4 w_n^{4i} + \cdots + a_{n-2} w_n^{(n-2)i}) \\
&\quad + w_n^i (a_1 + a_3 w_n^{2i} + a_5 w_n^{4i} + \cdots + a_{n-1} w_n^{(n-2)i}) \\
&= \left(a_0 + a_2 w_{n/2}^i + a_4 w_{n/2}^{2i} + \cdots + a_{n-2} w_{n/2}^{\frac{(n-2)i}{2}} \right) \\
&\quad + w_n^i \left(a_1 + a_3 w_{n/2}^i + a_5 w_{n/2}^{2i} + \cdots + a_{n-1} w_{n/2}^{\frac{(n-2)i}{2}} \right).
\end{aligned}$$

$$(3\text{-}167)$$

Define B_i and C_i as follows:

$$B_i = a_0 + a_2 w_{n/2}^i + a_4 w_{n/2}^{2i} + \cdots + a_{n-2} w_{n/2}^{\frac{(n-2)i}{2}} \qquad (3\text{-}168)$$

$$C_i = a_1 + a_3 w_{n/2}^i + a_5 w_{n/2}^{2i} + \cdots + a_{n-1} w_{n/2}^{\frac{(n-2)i}{2}}. \qquad (3\text{-}169)$$

Then:

$$A_i = B_i + w_n^i C_i. \qquad (3\text{-}170)$$

But $B_i(C_i)$ is the ith Fourier transform coefficient of $a_0, a_2, \ldots, a_{n-2}(a_1, a_3, \ldots, a_{n-1})$. Thus both B_i and C_i can be found recursively by using this approach.

It can also be proved that

$$A_{i+n/2} = B_i + w_n^{i+n/2} C_i. \tag{3-171}$$

For $n = 2$, $w_2 = \cos \pi - j \sin \pi = -1$ and the discrete Fourier transform is as follows:

$$A_0 = a_0 + w_2^0 a_1 = a_0 + a_1$$
$$A_1 = a_0 + w_2^1 a_1 = a_0 - a_1. \tag{3-172}$$

Fast Discrete Fourier Transform Algorithm Based on the Divide-and-Conquer Approach
Input: $a_0, a_1, \ldots, a_{n-1}, n = 2^p$.

Output: $A_i = \sum\limits_{k=0}^{n-1} a_k e^{-j\frac{2\pi ik}{n}}$, for $i = 0, 1, 2, \ldots, n - 1$.

Step 1: If $n = 2$ we have

$$A_0 = a_0 + a_1$$
$$A_1 = a_0 - a_1.$$

Step 2: Recursively find the coefficients of the discrete Fourier transform of $a_0, a_2, \ldots, a_{n-2}$ $(a_1, a_3, \ldots, a_{n-1})$. Let the coefficients be denoted as $B_0, B_1, \ldots, B_{n/2}$ (with $C_0, C_1, \ldots, C_{n/2}$).
Step 3: For $i = 0$ to $(n/2) - 1$:

$$A_i = B_i + w_n^i C_i$$
$$A_{i+n/2} = B_i + w_n^{i+n/2} C_i.$$

The complexity of the above algorithm is obviously of the order of $(n \log n)$.

Example 3-19
Let us consider $x(k) = \cos(2\pi k/4)$. We sample four points: $k = 0, 1, 2, 3$.

$$a_0 = \cos(0) = 1$$
$$a_1 = \cos(\pi/2) = 0$$
$$a_2 = \cos(\pi) = -1$$
$$a_3 = \cos(3\pi/2) = 0.$$

We first find the discrete Fourier transform of a_0 and a_2:

$$B_0 = a_0 + a_2 = 0$$
$$B_1 = a_0 - a_2 = 2.$$

We then find the discrete Fourier transform of a_1 and a_3:

$$C_0 = a_1 + a_3 = 0$$
$$C_1 = a_1 - a_3 = 0.$$

Finally

$$w_4^0 = 1$$
$$w_4^1 = e^{-j\frac{2\pi}{4}} = \cos\left(\frac{\pi}{2}\right) - j\sin\left(\frac{\pi}{2}\right) = -j$$
$$w_4^2 = (-j)^2 = -1$$
$$w_4^3 = (-j)^3 = j.$$

We merge the B_i and C_i to obtain the A_i.

$$A_0 = B_0 + w_4^0 C_0 = 0$$
$$A_1 = B_1 + w_4^1 C_1 = 2$$
$$A_2 = B_0 + w_4^2 C_0 = 0$$
$$A_3 = B_1 + w_4^3 C_1 = 2$$

What is the meaning of A_0, A_1, A_2 and A_3? **First of all, we note that among A_0 and A_1, only A_1 is not zero. Thus we conclude that there is a signal with frequency 1.**

We should also understand that using Equation (3-148), we can reconstruct a_0, a_1, a_2 and a_3 by using the A_i. Let us verify this by using the following calculation:

$$a_0 = \frac{1}{4}(A_0 + A_1 + A_2 + A_3) = \frac{1}{4}(2 + 2) = 1$$

$$a_1 = \frac{1}{4}\left(A_1 e^{j\frac{2\pi(1)(1)}{4}} + A_3 e^{j\frac{2\pi(1)(3)}{4}}\right)$$

$$= \frac{1}{4}\left(2e^{j\frac{\pi}{2}} + 2e^{j\frac{3\pi}{2}}\right)$$

$$= \frac{1}{4}(2j - 2j)$$

$$= 0$$

$$a_2 = \frac{1}{4}\left(A_1 e^{j\frac{2\pi(2)(1)}{4}} + A_3 e^{j\frac{2\pi(2)(3)}{4}}\right)$$

$$= \frac{1}{4}(2e^{j\pi} + 2e^{j3\pi})$$

$$= \frac{2}{4}(\cos(\pi) + j\sin(\pi) + \cos(3\pi) - j\sin(3\pi))$$

$$= \frac{2}{4}(-1 + 0 - 1 - 0)$$

$$= -1$$

$$a_3 = \frac{1}{4}\left(A_1 e^{j\frac{2\pi(3)(1)}{4}} + A_3 e^{j\frac{2\pi(3)(3)}{4}}\right)$$

$$= \frac{1}{4}\left(2e^{j\frac{3\pi}{2}} + 2e^{j\frac{9\pi}{2}}\right)$$

$$= \frac{2}{4}\left(\cos\left(\frac{3\pi}{2}\right) + j\sin\left(\frac{3\pi}{2}\right) + \cos\left(\frac{9\pi}{2}\right) + j\sin\left(\frac{9\pi}{2}\right)\right)$$

$$= \frac{2}{4}(0 - j + 0 + j)$$

$$= 0.$$

We can see that the a_i are correctly obtained.

Example 3-20
Suppose that we use

$$x(k) = \sin\left(\frac{2\pi}{4}k\right) = \sin\left(\frac{\pi}{2}k\right).$$

Again, we sample points at $k = 0, 1, 2$ and 3. In this case, $a_0 = 0, a_1 = 1, a_2 = 0$ and $a_3 = -1$. It can be easily seen that $B_0 = 0, B_1 = 0, C_0 = 0$ and $C_1 = 2$ and $A_0 = A_2 = 0, A_1 = -2j$ and $A_3 = 2j$. As in Example 3-18, among A_0 and A_1, only A_1 is not zero, indicating that there is a signal with frequency 1. The existence of the imaginary number j in A_1 means that the signal is a sine function. This will be explained in Section 3.10.

3.9 The Inverse Discrete Fourier Transform

The inverse discrete Fourier transform transforms the A_i back to a_i through the following equation:

$$a_i = \frac{1}{n}\sum_{k=0}^{n-1} A_k e^{j\frac{2\pi ik}{n}}.$$

The fast inverse discrete Fourier transform is almost exactly the same as the fast discrete Fourier transform. We merely have to replace w_n by w_n^{-1} and add $1/n$ in appropriate places.

The following is the fast inverse discrete Fourier transform algorithm.

Fast Inverse Discrete Fourier Transform Algorithm
Input: $A_0, A_1, \cdots, A_{n-1}, n = 2^p$.
Output: $a_i = \frac{1}{n}\sum_{k=0}^{n-1} A_k e^{j\frac{2\pi ik}{n}}$ for $i = 0, 1, 2, \cdots, n-1$.

Step 1: If $n = 2$:

$$a_0 = \frac{1}{2}(A_0 + A_1)$$

$$a_1 = \frac{1}{2}(A_0 - A_1).$$

Step 2: Recursively find the coefficients of the inverse discrete Fourier transform of $A_0, A_2, \ldots, A_{n-2}(A_1, A_3, \ldots, A_{n-1})$. Let the coefficients be denoted as $B_0, B_1, \ldots, B_{n/_2-1}$ $(C_0, C_1, \ldots, C_{n/_2-1})$.

Step 3: For $i = 0$ to $i = \dfrac{n}{2} - 1$:

$$a_i = \frac{1}{n}(B_i + w_n^{-i}C_i)$$

$$a_{i+n/2} = \frac{1}{n}(B_i + w_n^{-(i+n/2)}C_i).$$

Example 3-21

Take the result of Example 3-18. We have $A_0 = A_2 = 0, A_1 = A_3 = 2$. We will have $B_0 = B_1 = C_1 = 0$, and $C_0 = 3$. We thus have

$$a_0 = \frac{1}{4}(B_0 + w_4^0 C_0) = \frac{1}{4}C_0 = \frac{1}{4}(4) = 1$$

$$a_1 = \frac{1}{4}(B_1 + w_4^{-1}C_1) = 0$$

$$a_2 = \frac{1}{4}(B_0 + w_4^{-2}C_0) = \frac{1}{4}((-j)^{-2}C_0) = \frac{1}{4}(-4) = -1$$

$$a_3 = \frac{1}{4}(B_1 + w_4^{-3}C_1) = 0$$

which is correct.

Example 3-22

Take the result of Example 3-19. In this case, we have $A_0 = A_2 = 0$, $A_1 = -2j$ and $A_3 = 2j$. This means that $B_0 = B_1 = C_0 = 0$ and $C_1 = -4j$.

Finally, we have

$$a_0 = \frac{1}{4}(B_0 + w_4^0 C_0) = 0$$

$$a_1 = \frac{1}{4}(B_1 + w_4^{-1}C_1) = \frac{1}{4}(0 + w_4^{-1}(-4j)) = (-j)^{-1}(-j) = 1$$

$$a_2 = \frac{1}{4}(B_0 + w_4^{-2}C_0) = 0$$

$$a_3 = \frac{1}{4}(B_1 + w_4^{-3}C_1) = \frac{1}{4}(0 + w_4^{-3}(-4j)) = (-j)^{-3}(-j) = -1$$

which is as expected.

3.10 Physical Meaning of the Discrete Fourier Transform

In the previous sections, we have discussed many mathematical properties of the discrete Fourier transform. In this section, we shall try to give the reader more feeling about the physical meaning of this transform.

In practice, we are not given a mathematical definition of a function $x(t)$ and try to find a Fourier transform of this function. Instead, we are given a function through sampling. Let the time period be T. Suppose we sample n times. Let $T_s = T/n$. That is, we sample this function and obtain $a_0, a_1, \ldots, a_{n-1}$, where $a_k = x(kT_s)$. Without risking ambiguity, we may denote $x(k) = x(kT_s)$ for simplicity. In the following, we may assume that $T_s = 1$. Thus, our function $x(t)$ is actually characterized by the following relation:

$$x(0) = a_0$$
$$x(1) = a_1$$
$$\vdots$$
$$x(n-1) = a_{n-1}.$$

From Equation (3-62), our problem is to find the A_i such that

$$x(k) = x(kT_s) = \frac{1}{n}\sum_{i=0}^{n-1} A_i e^{\frac{j2\pi ik}{n}}, \text{ for } k = 0, 1, \cdots, n-1. \tag{3-173}$$

Note that

$$e^{\frac{j2\pi ik}{n}} = \cos\left(\frac{2\pi i}{n}k\right) + j\sin\left(\frac{2\pi i}{n}k\right).$$

It is thus quite natural for a reader to have the following questions:

1. What is the physical meaning of the imaginary part of the signal, namely $j\sin\{(2\pi i/n)k\}$?
2. When A_i is a complex number, i.e. $A_i = a_i + jb_i$, what is the physical meaning of it?

In general, in the discrete Fourier transform of any real signal, for $1 \leq i \leq (n/2) - 1$, if $A_i = a + jb$, then $A_{n-i} = a - jb$. This is rather easy to show.

Recalling that

$$w_n = e^{-j\frac{2\pi}{n}},$$

from Equation (3.6-1) we have

$$A_i = a_0 + a_1 w_n^i + a_2 w_n^{2i} + \ldots + a_n^{(n-1)i}$$
$$A_{n-i} = a_0 + a_1 w_n^{n-i} + a_2 w_n^{2(n-i)} + \ldots + a_n^{(n-1)(n-i)} \tag{3-174}$$
$$= a_0 + a_1 w_n^{-i} + a_2 w_n^{-2i} + \ldots + a_n^{-(n-1)i}.$$

Since

$$w_n^i = e^{-j\frac{2\pi i}{n}} = \cos\left(\frac{2\pi i}{n}\right) - j\sin\left(\frac{2\pi i}{n}\right), w_n^{-i} = e^{j\frac{2\pi i}{n}} = \cos\left(\frac{2\pi i}{n}\right) + j\sin\left(\frac{2\pi i}{n}\right) \text{ and } a_i$$

are real numbers, if $A_i = a + jb$, then $A_{n-i} = a - jb$.

Note that, from Equation (3-173):

$$x(k) = \frac{1}{n} \sum_{i=0}^{n-1} A_i e^{j\frac{2\pi ik}{n}}, \text{ for } k = 0, 1, 2, \cdots, n-1. \tag{3-175}$$

Let $A_i = a + jb = re^{j\theta}$, where $\theta = \tan^{-1}(b/a)$. Then $A_{n-i} = a - jb = re^{-j\theta}$. By using the fact $e^{2\pi i} = 1$, we can easily see that

$$A_i e^{j\frac{2\pi ik}{n}} + A_{n-i} e^{j\frac{2\pi k(n-i)}{n}}$$
$$= A_i e^{j\frac{2\pi ik}{n}} + A_{n-i} e^{j2\pi k} e^{-j\frac{2\pi ik}{n}}. \tag{3-176}$$

Since $e^{j2\pi k} = 1$ for $k = 0, 1, \ldots, \text{n-1}$, we have

$$A_i e^{j\frac{2\pi ik}{n}} + A_{n-i} e^{j\frac{2\pi k(n-i)}{n}}$$
$$= A_i e^{j\frac{2\pi ik}{n}} + \overline{A_i} e^{-j\frac{2\pi ik}{n}}$$
$$= re^{j\theta_i} e^{j\frac{2\pi ik}{n}} + re^{-j\theta_i} e^{-j\frac{2\pi ik}{n}} \quad \text{for } k = 0, 1, 2, \cdots, n-1.$$
$$= re^{j\left(\theta_i + \frac{2\pi ik}{n}\right)} + re^{-j\left(\theta_i + \frac{2\pi ik}{n}\right)}$$
$$= 2r \cos\left(\frac{2\pi ik}{n} + \theta_i\right). \tag{3-177}$$

So we can see that, actually, the discrete Fourier transform contains a sequence of cosine functions with possibly phase shifts. That is, the Fourier transform of any function can be expressed as follows:

$$x(k) = \frac{1}{n}\left(A_0 + A_{n/2} e^{j\pi k} + 2 \sum_{i=1}^{n/2-1} |A_i| \cos\left(\frac{2\pi ik}{n} + \theta_i\right)\right), \text{ for } k = 0, 1, 2, \cdots, n-1. \tag{3-178}$$

From the above equation, it can be seen that, by using the discrete Fourier transform, we may approximate a function $x(t)$ as a superposition of cosine functions. The following observations are in order.

1. *The job of the discrete Fourier transform is to find A_k from a_i where $a_k = x(k) = x(kT_s)$ for $i = 0, 1, \cdots, n-1$. Thus we may say that the discrete Fourier transform is an analysis tool. It transforms $x(t)$ from the time domain to the frequency domain and enables us to find out what frequencies exist in $x(t)$.*
2. *The job of the inverse discrete Fourier transform is to find a_i from A_i. Thus we may say that the inverse discrete Fourier transform is a reconstruction tool. It transforms $x(t)$ from the frequency domain to the time domain and recovers the original function.*
3. *The reconstruction is not complete because Equation (3-173) is valid only for those n points, i.e. $x(0), x(T_s), x(2T_s), \ldots, x((n-1)T_s)$ The larger n is, the more accurately Equation (3-173) represents $x(t)$.*

4. *Since it is not easy to have a very large n, we may use some interpolation method to reconstruct* $x(t)$.
5. *Among* the A_i, *for* $i = 0$ *to* $n - 1$, *we only have to pay attention to* A_0, $A_{n/2}$ *and the* A_i *for* $i = 1$ *to* $(n/2) - 1$, *as seen in Equation (3-178).*
6. A_i *and* A_{n-i} *make contributions to the magnitude and phase shift of* $\cos(2\pi i k/n)$, *again as seen in Equation (3-178).*

Perhaps the reader may legitimately ask the following question: Since the discrete Fourier transform effectively transforms $x(t)$ to a set of cosine functions, is there a cosine transform which transforms a function $x(t)$ to a superposition of cosine functions directly? Yes, there is a cosine transform. The unfortunate thing is that there is no elegant fast discrete cosine transform as we have in the discrete Fourier transform case. This is why the discrete Fourier transform is still used.

Example 3-23
Let us go back to Example 3-18 where $x(k) = \cos(2\pi k/4)$. In this case, $A_0 = A_2 = 0$ and $A_1 = A_3 = 2$. We can see that

$$x(k) = \frac{1}{4}\left(2e^{j\frac{2\pi k}{4}} + 2e^{j\frac{2\pi(3)k}{4}}\right)$$
$$= \frac{1}{2}\left(e^{j\frac{\pi k}{2}} + e^{j\frac{3\pi k}{2}}\right)$$
$$= \frac{1}{2}\left(e^{j\frac{\pi k}{2}} + e^{j(2\pi k - \frac{\pi k}{2})}\right)$$
$$= \frac{1}{2}\left(e^{j\frac{\pi k}{2}} + e^{j2\pi k}e^{-j\frac{\pi k}{2}}\right).$$

Since $e^{j2\pi k} = 1$ for $k = 0, 1, 2, 3$, we have

$$x(k) = \frac{1}{2}\left(e^{j\frac{\pi k}{2}} + e^{-j\frac{\pi k}{2}}\right) = \cos\left(\frac{\pi k}{2}\right) \text{ for } k = 0, 1, 2, 3.$$

Actually, we obtain the same result by using Equation (3-178). In this case:

$$A_0 = A_{n/2} = A_2 = 0$$
$$|A_1| = 2, \ \theta_1 = 0.$$

Therefore, according to Equation (3-178), we have

$$x(k) = \frac{1}{4}\left(2 \cdot 2 \cos\left(\frac{2\pi k}{4}\right)\right)$$
$$= \cos\left(\frac{\pi k}{2}\right), \quad \text{for} \quad k = 0, 1, 2, 3.$$

Assume that $T_s = 1$. *We conclude that although we cannot say that* $x(t) = \cos(\pi t/2)$ *for all t, we can say that* $x(t) = \cos(\pi t/2)$ *for* $t = 0, 1, 2, 3$.

In the above case, the non-zero A_i are real numbers and the function is found to be a cosine function. Let us now consider Example 3-20. In this case, the non-zero A_i are imaginary numbers: $A_1 = -2j$ and $A_3 = 2j$.

$$
\begin{aligned}
x(k) &= \frac{1}{4}\left((-2j)e^{j\frac{2\pi k}{4}} + 2je^{j\frac{2\pi(3)k}{4}}\right) \\
&= \frac{1}{4}(2j)\left(-e^{j\frac{\pi k}{2}} + e^{j\frac{3\pi k}{2}}\right) \\
&= \frac{1}{2}j\left(-e^{j\frac{\pi k}{2}} + e^{j\frac{3\pi k}{2}}\right) \\
&= \frac{1}{2}j\left(-e^{j\frac{\pi k}{2}} + e^{j2\pi k}e^{-j\frac{\pi k}{2}}\right) \\
&= \frac{1}{2}j\left(-e^{j\frac{\pi k}{2}} + e^{-j\frac{\pi k}{2}}\right) \\
&= \sin\frac{\pi k}{2} \quad \text{for} \quad k = 0, 1, 2, 3.
\end{aligned}
$$

Again, we cannot say that

$$
x(t) = \frac{1}{2}j\left(-e^{j\frac{\pi t}{2}} + e^{j\frac{3\pi t}{2}}\right) = \sin\left(\frac{\pi t}{2}\right)
$$

for all t. But, by using the techniques we used above, we can prove that

$$
x(t) = \frac{1}{2}j\left(-e^{j\frac{\pi t}{2}} + e^{j\frac{3\pi t}{2}}\right) = \sin\left(\frac{\pi t}{2}\right)
$$

for t = 0, 1, 2, 3.

Example 3-24

Consider $x(k) = \cos\{(2\pi k/4) + (\pi/4)\}$. In this case, we have

$$
a_0 = a_3 = \sqrt{2}/2 \text{ and } a_1 = a_2 = -\sqrt{2}/2.
$$

Thus, from Equations (3-154) to (3-157), we can prove that

$$
\begin{aligned}
A_0 &= 0 \\
A_1 &= \sqrt{2}(1+j) = 2e^{j\frac{\pi}{4}} \\
A_2 &= 0 \\
A_3 &= \sqrt{2}(1-j) = 2e^{-j\frac{\pi}{4}}.
\end{aligned}
$$

By using the Fourier transform equation for $n = 4$, we have

$$
\begin{aligned}
x(k) &= \frac{1}{2}\left(e^{j\left(\frac{\pi k}{2}+\frac{\pi}{4}\right)} + e^{j\left(\frac{3\pi k}{2}-\frac{\pi}{4}\right)}\right) \\
&= \frac{1}{2}\left(e^{j\left(\frac{\pi k}{2}+\frac{\pi}{4}\right)} + e^{j\left(2\pi k-\frac{\pi k}{2}-\frac{\pi}{4}\right)}\right)
\end{aligned}
$$

$$= \frac{1}{2}\left(e^{j\left(\frac{\pi k}{2}+\frac{\pi}{4}\right)} + e^{j2\pi k}e^{-j\left(\frac{\pi k}{2}+\frac{\pi}{4}\right)}\right), \text{for } k = 0, 1, 2, 3.$$

Since $e^{j2\pi ik} = 1$ for $k = 0, 1, 2, 3$, we have

$$x(k) = \frac{1}{2}\left(e^{j\left(\frac{\pi k}{2}+\frac{\pi}{4}\right)} + e^{-j\left(\frac{\pi k}{2}+\frac{\pi}{4}\right)}\right)$$
$$= \frac{1}{2}(2)\cos\left(\frac{\pi k}{2} + \frac{\pi}{4}\right)$$
$$= \cos\left(\frac{\pi k}{2} + \frac{\pi}{4}\right), \quad \text{for} \quad k = 0, 1, 2, 3.$$

Another way to obtain the result is to use Equation (3-178). In this case:

$$A_0 = A_{n/2} = A_2 = 0.$$
$$|A_1| = 2, \theta_1 = \frac{\pi}{4}.$$

Substituting those into Equation (3-178), we obtain

$$x(k) = \frac{1}{4}\left(2 \cdot 2 \cos\left(\frac{2\pi k}{4} + \frac{\pi}{4}\right)\right)$$
$$= \cos\left(\frac{\pi k}{2} + \frac{\pi}{4}\right), \text{for } k = 0, 1, 2, 3.$$

Again, although we cannot prove that $x(t) = \cos\{(\pi t/2) + (\pi/4)\}$ for all t, we can prove that it is true for $t = 0, 1, 2$ and 3.

Example 3-25
Consider $x(k) = k/4$ with $n = 4$, so that

$$a_0 = 0, \quad a_1 = \frac{1}{4}, \quad a_2 = \frac{2}{4} = \frac{1}{2}, \quad a_3 = \frac{3}{4}.$$

Using Equations (3-154) to (3-157) in Section 3.8, we have

$$A_0 = a_0 + a_1 + a_2 + a_3 = \frac{0}{4} + \frac{1}{4} + \frac{2}{4} + \frac{3}{4} = \frac{6}{4} = \frac{3}{2}.$$
$$A_1 = (a_0 - a_2) - j(a_1 - a_3)$$
$$= \left(0 - \frac{1}{2}\right) - j\left(\frac{1}{4} - \frac{3}{4}\right)$$
$$= -\frac{1}{2} + j\frac{1}{2}$$
$$= \frac{1}{2}(-1 + j).$$

$$A_2 = a_0 - a_1 + a_2 - a_3$$

$$= 0 - \frac{1}{4} + \frac{1}{2} - \frac{3}{4}$$

$$= -\frac{1}{2}.$$

$$A_3 = \overline{A_1} = \frac{1}{2}(-1 - j).$$

$$|A_1| = \frac{1}{\sqrt{2}}, \quad \theta_1 = \tan^{-1}\left(\frac{1}{-1}\right) = \frac{3\pi}{4}.$$

Thus, according to Equation (3-178), we have

$$x(k) = \frac{1}{4}\left(\frac{3}{2} - \frac{1}{2}e^{j\pi k} + \frac{2}{\sqrt{2}}\cos\left(\frac{2\pi k}{4} + \frac{3\pi}{4}\right)\right), \quad \text{for} \quad k = 0, 1, 2, 3.$$

Let us check this by considering some cases.

Case 1: $k = 0$:

$$x(0) = \frac{1}{4}\left(\frac{3}{2} - \frac{1}{2}e^{j0\pi} + \frac{2}{\sqrt{2}}\cos\left(\frac{2\pi(0)}{4} + \frac{3\pi}{4}\right)\right)$$

$$= \frac{1}{4}\left(\frac{3}{2} - \frac{1}{2} + \frac{2}{\sqrt{2}}\left(-\frac{\sqrt{2}}{2}\right)\right)$$

$$= \frac{1}{4}\left(\frac{3}{2} - \frac{1}{2} - 1\right)$$

$$= 0$$

$$= a_0.$$

Case 2: $k = 1$:

$$x(1) = \frac{1}{4}\left(\frac{3}{2} - \frac{1}{2}e^{j\pi(1)} + \frac{2}{\sqrt{2}}\cos\left(\frac{2\pi(1)}{4} + \frac{3\pi}{4}\right)\right)$$

$$= \frac{1}{4}\left(\frac{3}{2} - \frac{1}{2}(-1) + \frac{2}{\sqrt{2}}\cos\left(\frac{5\pi}{4}\right)\right)$$

$$= \frac{1}{4}\left(\frac{3}{2} + \frac{1}{2} + \frac{2}{\sqrt{2}}\left(-\frac{\sqrt{2}}{2}\right)\right)$$

$$= \frac{1}{4}\left(\frac{3}{2} + \frac{1}{2} - 1\right)$$

$$= \frac{1}{4}$$

$$= a_1.$$

Case 3: $k = 2$:

$$x(2) = \frac{1}{4}\left(\frac{3}{2} - \frac{1}{2}e^{j\pi(2)} + \frac{2}{\sqrt{2}}\cos\left(\frac{\pi(2)}{2} + \frac{3\pi}{4}\right)\right)$$

$$= \frac{1}{4}\left(\frac{3}{2} - \frac{1}{2} + \frac{2}{\sqrt{2}}\cos\left(\frac{7\pi}{4}\right)\right)$$

$$= \frac{1}{4}\left(\frac{3}{2} - \frac{1}{2} + \frac{2}{\sqrt{2}}\frac{\sqrt{2}}{2}\right)$$

$$= \frac{1}{4}\left(\frac{3}{2} - \frac{1}{2} + 1\right)$$

$$= \frac{1}{4}(2)$$

$$= \frac{1}{2}$$

$$= a_2.$$

Case 4: $k = 3$:

$$x(3) = \frac{1}{4}\left(\frac{3}{2} - \frac{1}{2}e^{j3\pi} + \frac{2}{\sqrt{2}}\cos\left(\frac{\pi(3)}{2} + \frac{3\pi}{4}\right)\right)$$

$$= \frac{1}{4}\left(\frac{3}{2} - \frac{1}{2}(-1) + \frac{2}{\sqrt{2}}\cos\left(\frac{9\pi}{4}\right)\right)$$

$$= \frac{1}{4}\left(\frac{3}{2} + \frac{1}{2} + \frac{2}{\sqrt{2}}\frac{\sqrt{2}}{2}\right)$$

$$= \frac{1}{4}\left(\frac{3}{2} + \frac{1}{2} + 1\right)$$

$$= \frac{1}{4}(3)$$

$$= \frac{3}{4}$$

$$= a_3.$$

Thus, we have successfully used the discrete Fourier transform to approximate $x(t)$. It should be noted that this result is valid only for $t = 0, 1, 2, 3$.

In general, let $A_i = a + jb$. We have the following rules:

- **Rule 1**. If $b = 0$, there is a signal which is $a\cos(2\pi ik/n)$.
- **Rule 2**. If $a = 0$, there is a signal which is $b\sin(2\pi k/n)$.
- **Rule 3**. If neither of a or b is zero, there is a signal which is $\sqrt{a^2 + b^2}\cos\{(2\pi ik/n) + \theta_i\}$, where $\theta_i = \tan^{-1}(b/a)$.

Since the complex representation of an A_i indicates the existence of a cosine function with a certain frequency shift, can we ignore the phase shift and only pay attention to the magnitude, namely $\sqrt{a^2 + b^2}$? We certainly can. Suppose we have obtained the A_i. We transform all of the A_i by ignoring the phase-shift form by preserving only its magnitude. Then we perform the inverse Fourier transform on this set of transformed A_i. The result will be equal to a sum of cosine functions with the original frequencies with their original phase shifts ignored. Let us consider the result of Example 3-23. The function considered is a cosine function with a phase shift of $\pi/4$. We also note that $A_1 = \sqrt{2}(1 + j1)$ and $A_3 = \sqrt{2}(1 - j1)$. If we ignore the phase part, we will have $A_0 = A_2 = 0$, $A_1 = 2$ and $A_3 = 2$. We now have the same result as that in Example 3-18 where a cosine function without any phase shift was considered. This means that if we ignore the phase shift, we will produce a signal without the phase shift. But the frequency of the signal is preserved.

Let us see how we can apply the discrete Fourier transform to signals. First of all, after performing a discrete Fourier transform, we can see some of the A_i have rather large amplitudes. We preserve the amplitudes of these A_i and ignore all of the other A_i with small amplitudes. Thus we have effectively conducted a data compression in a certain sense. We may only transmit these A_i. On the other side, after receiving these A_i, we can perform an inverse Fourier transform on these A_i. The result is, of course, not exactly the same as the original signal, but the result should be quite close to the original. For the human voice, this kind of data compression usually works as we are not that particular about the actual sound of a particular voice.

Note that in the Fourier transform, each A_i which is not zero corresponds to a signal with frequency i. Let us consider the signal in Figure 3-27. A Fourier transform spectrum is shown in Figure 3-28.

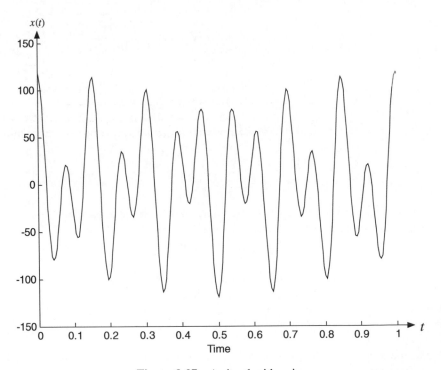

Figure 3-27   A signal with noise

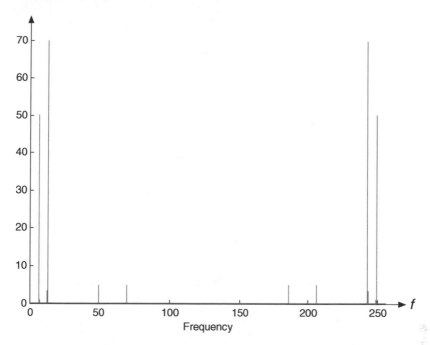

Figure 3-28   The discrete Fourier transform of the signal in Figure 3-27

In this case, we set $n = 256$. In Figure 3-28, we first note that there are eight peaks. But they appear in pairs, since the original signal is real. Therefore, only two of them are significant. We should only pay attention to half of the spectrum, namely from 0 to $n/2 = 128$. In this part, there are four peaks, located at 7, 13, 50 and 70. This indicates that the signal in Figure 3-27 contains four sinusoidal signals with frequencies 7, 13, 50 and 70. The magnitudes of signals with frequencies 50 and 70 are rather small, as compared with those of the signals with frequencies 7 and 13. As for the part of the spectrum from $n/2 = 128$ to $n = 256$, the peaks are all symmetrical to $n/2$. For instance, for the peak at 13, there is an identical peak at $n - 13 = 256 - 13 = 243$. Similarly, for the small peak at 50, there is also a peak at $n - 50 = 256 - 50 = 206$.

Example 3-26
Let us consider Figure 3-27 again. We have a signal which obviously contains some kind of noise. Figure 3-28 shows the result of the application of the discrete Fourier transform. There are several peaks with very low magnitudes. Figure 3-29 shows the inverse discrete Fourier transform without considering these low-magnitude peaks. We can now see that noise is successfully eliminated.

Example 3-27
Figure 3-30 shows a period of some music signal which lasts one second. In Figure 3-31 we show a discrete Fourier thow a discrete Fourier transform of the signal.

In this analysis, n was set to be $2^{14} = 16,384$. Therefore the sampling interval is $T_s = 1/16384$ (seconds). The sampling interval in the frequency domain is $f_0 = 1/T$.

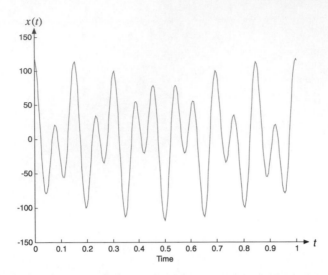

Figure 3-29   The signal obtained by ignoring the low-amplitude peaks

Since the signal has length of $T = 1$ (seconds), $f_0 = 1$ hertz. Therefore, the resolution in the frequency domain is 1 hertz. As we understand, the discrete Fourier transform spectrum is symmetrical to $n/2$. Thus only those frequency components lower than 8192 hertz are shown. In other words, our analysis cannot determine whether the signal contains signals with

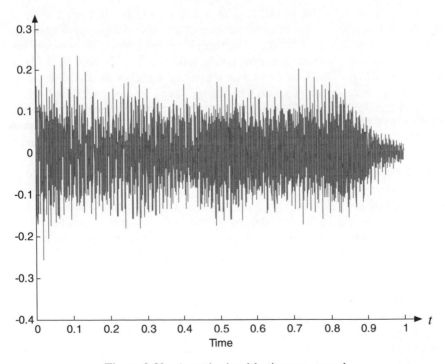

Figure 3-30   A music signal lasting one second

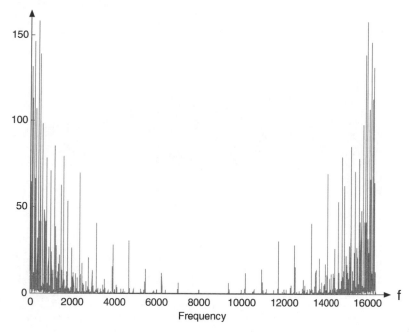

Figure 3-31   Discrete Fourier transform spectrum of the signal in Figure 3-29

frequencies higher than 8192 hertz. The reader should note that the discrete Fourier transform coefficients are complex numbers. In Figure 3-31, we only show their magnitudes.

We then performed an inverse discrete Fourier transform and the result is shown in Figure 3-32. As one can see, the inverse discrete Fourier transform successfully recovers the original signal.

Among the frequency components shown in Figure 3-31, some are of low amplitude. We first eliminated those discrete Fourier transform coefficients whose amplitudes were smaller than 5. Then we performed an inverse discrete Fourier transform with the remaining coefficients. The result is shown in Figure 3-33.

We further eliminated those coefficients whose amplitudes were smaller than 10 and performed an inverse discrete Fourier transform on the remaining coefficients. The result is shown in Figure 3-34. As can be seen, it is acceptable to eliminate those frequency components whose amplitudes are not large.

Let us take a look at Fig. 3-31 again. Note that the spectrum of the discrete Fourier transform of any signal is symmetrical with respect to $n/2$. Thus the highest frequency that the spectrum can exhibit will be less than $n/2$. Let us assume that the highest frequency which a signal contains is f and the sampling rate is n. Then, in order to be effective, n must be larger than $2f$. The highest significant frequency in a human voice is in general less than 8000 hertz. This is why we let the sampling rate $1/T_s = n$ be around 16,000 hertz. Furthermore, the time we used for our human voice signal is one second. Suppose the time period of our signal is m seconds and the highest frequency is f. The sampling rate then has to be $2mf$.

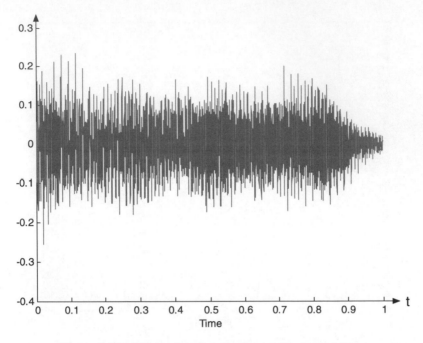

Figure 3-32 Result of the inverse discrete Fourier transform

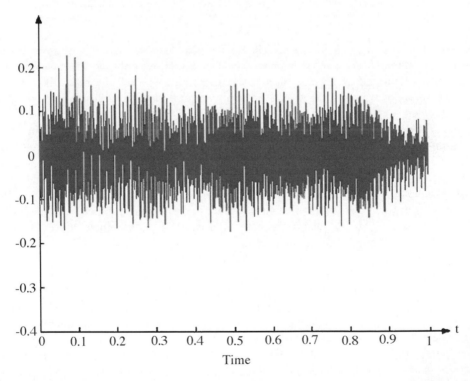

Figure 3-33 Inverse discrete Fourier transform after eliminating frequency components whose amplitudes are smaller than 5

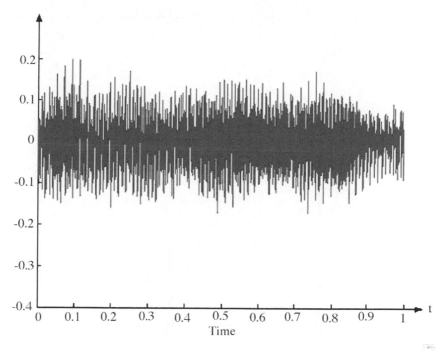

Figure 3-34 Inverse discrete Fourier transform after eliminating frequency components whose amplitudes are smaller than 10

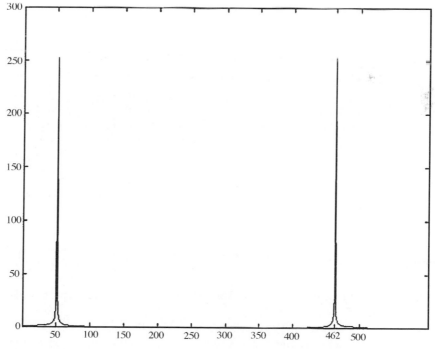

Figure 3-35 Discrete Fourier transform of $\cos\{2\pi(50)t\}$ where the sampling time is one second

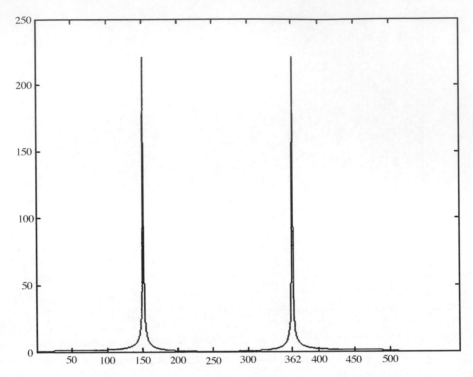

Figure 3-36 Discrete Fourier transform of $\cos\{2\pi(50)t\}$ where the sampling time is 3 seconds

Finally, note that each time we sampled a function, the time period was set to be one second. If it is, say, 3 seconds, the frequencies will be $3f$. In Figure 3-35, we show the discrete Fourier transform of $\cos\{2\pi(50)t/n)\}$, where $n = 512$ and the time period of sampling was one second. The peak is at $f = 50$. We then set the time period to be 3 seconds. As shown in Figure 3-36, the peak is shifted to 150. This shows that a normalization of frequencies has to be done if the time period of sampling is more than one second.

Further Reading

- For wide coverage of time-domain and frequency-domain analysis of both discrete-time and continuous-time systems, see [OWY83].
- [P62] and [B65] provide in-depth analysis of the Fourier series and transform techniques.

Exercises

3.1   Consider the function in Figure 3-12(a). Show that its Fourier series is as shown in Equation (3-35).

3.2   The function in Figure 3-12(a) may be considered as the function in Fig. 3-9 moved $T/4$ to the left with its magnitude reduced to half. Let $f_1(t)$ be the Fourier series of the function in Figure 3-9 and $f_2(t)$ be that of the function in Figurre 3-12(a). Show that $f_2(t) = (1/2)f_1\{t + (T/4)\}$.

3.3 Complete the following equations:

$$e^{j\theta} =$$

$$e^{-j\theta} =$$

$$e^{j\theta} + e^{-j\theta} =$$

$$e^{j\theta} - e^{-j\theta} =$$

$$\cos 2\theta = \frac{1}{2}(\quad)$$

$$\sin\theta = \frac{1}{2j}(\quad)$$

3.4 Consider Example 3-3. Obtain its Fourier series in exponential expressions by using Equations (3-20) to (3-22) and $X_k = (a_k/2) + (b_k/2j)$. Check your answer with Equation (3-68).

3.5 Consider Equation (3-64) again. Function $x(t)$ is a real function. Yet in Equation (3-64), imaginary terms exist. Besides, negative frequencies also exist. How do you explain this?

3.6 Consider the term $X_{-k}e^{-j2\pi kf_0t} + X_k e^{j2\pi kf_0t}$. Do you understand that the above term is actually a cosine function? Explain.

3.7 Find the Fourier transform of the following function.

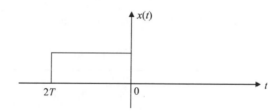

Compare the result with Equation (3-102). Explain.

3.8 Let $F(x(t)) = X(f)$. Complete the following:

$$F(x(t - t_0)) =$$

$$F(e^{j2\pi\alpha t}x(t)) =$$

$$F(\cos(2\pi\alpha t)x(t)) =$$

$$F^{-1}(e^{-j2\pi ft_0}X(f)) =$$

$$F^{-1}(X(f - \alpha)) =$$

$$F^{-1}(X(f + \alpha) + X(f - \alpha)) =$$

3.9 State the definition of convolution.

3.10 Let $F(x(t)) = X(f)$ and $F(y(t)) = Y(f)$. Let $z(t) = x(t)y(t)$. Show how we can find $F(z(t))$ by using the concept of convolution.

3.11   Consider the following two functions $x(u)$ and $y(u)$.

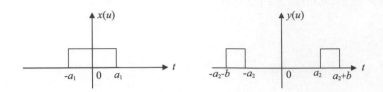

Explain why the convolution of $x(u)$ and $y(u)$ looks like the following by showing how the convolution is found step by step. Find x_1, x_2, y_1 and y_2.

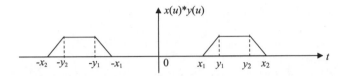

3.12   Complete the following:

$$F(\delta(t)) =$$
$$F(1) =$$

Explain how the above results can be reasoned out by using the result in Example 3-12.

3.13   Complete the following

$$F(e^{j2\pi\alpha t}) =$$
$$F(\cos(2\pi\alpha t)) =$$
$$F(\sin(2\pi\alpha t)) =$$

3.14   Let $X(f) = F(x(t))$ be as shown below. Let $z(t) = x(t)\cos(2\pi\alpha t)$.

$$Z(t) = x(t)y(k) \xrightarrow[F^{-1}]{F} Z(f) = X(f) * Y(f)$$

$$F(z(t)) = F(x(t)\cos(2\pi\alpha t))$$

$$= F(x(t) * \frac{1}{2}(\delta(f-a) + \delta(f+a))$$

$$= \frac{1}{2}(X(f-\alpha) + X(f+\alpha))$$

Find $F(z(t))$ by using the convolution theorem of Fourier transform.

3.15   (a) Let $x(t) = \cos(2\pi t/4)$. Let $a(i) = x(i)$ for $i = 0, 1, 2, 3$, Find the A_i of the discrete Fourier transform for $i = 0, 1, 2, 3$.
       (b) Show that the equation of $x(t)$ in terms of A_i will recover $x(t)$ for $i = 0, 1, 2, 3$.

3.16   Do the same thing as above, except now for $x(t) = 1 + \sin(2\pi/4)$.

3.17 Do the same as in Exercise 3.16, except now for $x(t) = \cos\{(2\pi t/4) + (\pi/3)\}$.

3.18 Do the same as in Exercise 3.16, except now for $x(t) = \cos\{(2\pi t/4) + (\pi/6)\}$.

3.19 Let $x(t) = t/4$ and $n = 4$. Use the discrete Fourier transform to express $x(t)$ as a series of cosine functions. Verify the result by substituting $x = 0, 1, 2, 3$.

3.20 Do the same as in Exercise 3.20, except for $x(t) = t^2/4$.

3.21 Prove that formulas in Equation (3-29) are valid for the signal in Figure 3-10.

3.22 Prove that Equations (3-43) to (3-45) are valid for the signal in Figure 3-13(c).

3.23 For Example 3-15, obtain the same result by sliding $x(u)$ to the left and keeping $y(u)$ fixed. You may consider Figure 3-21.

4

Analog Modulation Techniques

In communications engineering, one is concerned with the transmission of certain information from one point to another. Some basic knowledge is quite important.

1. There are two types of communication system, wired and wireless.
2. The media for the information passing from one point to another are different for different kinds of communication system. For example, some systems use a 'twisted pair' for voice transmission, and cable TV uses coaxial cable as the transmission medium. Some data network systems use optical fiber links. Free space is the transmission medium for wireless transmission systems, where communication between two points relies on the propagation of electromagnetic waves. Examples of wireless communication are radio and television broadcasting, cordless telephones, pagers, mobile phone systems, wireless local networks (WLAN), etc.
3. The information can be analog or digital. Human voice or music is analog while texts or other types of data are often digital.
4. An analog signal can be transmitted directly over wired channels. If the digital information is encoded by pulses, it can be transmitted directly over wired channels. However, for wireless systems, digital encoded baseband pulses cannot be transmitted directly, except when the digital information is encoded by very narrow pulses. The systems that use very narrow pulses to transmit digital information are called ultra-wideband (UWB) systems. They work because the narrow pulses contain large amounts of high-frequency components. Therefore, the narrow pulses can still be transmitted over wireless channels. In general, to be practically feasible, only high-frequency signals can be sent over wireless channels because of the properties of electromagnetic waves.
5. For wireless systems, antennas must be used.
6. For both systems, different signals may be sent through one channel. For example, TV broadcasts from 60 different TV stations may be transmitted through the same cable. All of the original signals are called *baseband* signals. Some method must be used to distinguish these baseband signals.

This chapter describes various analog modulation techniques. Analog modulation translates a baseband analog message signal to a bandpass signal at a frequency band that is

Communications Engineering.   R.C.T. Lee, Mao-Ching Chiu and Jung-Shan Lin
© 2007 John Wiley & Sons (Asia) Pte Ltd

higher than that of the baseband message signal. The resulting signal is called the **modulated signal** and the baseband message signal is called the **modulating** signal. We shall explain the meaning of 'bandpass' later.

One may ask: When do we need modulation techniques? The answer depends on the transmission medium. First of all, it is very common for multiple message signals to be transmitted simultaneously through the same channel, as we mentioned above. These baseband signals are of roughly the same low frequency. Under this circumstance, some modulation techniques are needed so that different messages can be modulated into different frequency bands. In other words, after modulation, different signals are translated into different frequency bands. These signals are then transmitted through the same channel. At the receiving end, the receiver must be capable of separating different modulated signals and then demodulating each message signal separately.

Consider radio broadcasting in any area. There will be several stations broadcasting essentially music and voice. Thus the baseband signals are all over the low-frequency band. To differentiate these signals, analog modulation can be used. After this kind of modulation, different stations are identified with different frequencies. We can now listen to radio broadcasts essentially because of this modulation. The cable TV system is another typical case where analog modulation is necessary.

For wireless transmission, there is another strong reason why modulation techniques must be used. Since the medium for wireless transmission is free space, the baseband message signal is not feasible for direct transmission. This is because the antenna to transmit a signal with frequency f needs to be an antenna with length $\lambda/2$, where λ is the wavelength of the signal. Unfortunately, limited by the scope of this book, we cannot elaborate the concept of wavelength. At present, we merely say that the wavelength λ of a signal is inversely proportional to its frequency f, so their relationship is expressed as $\lambda = v/f$, where v is the speed of light which is 3×10^8 m/s. By performing Fourier transform analysis on a human voice, we will find that it contains signals concentrated within 3000 hertz. The minimum wavelength of the voice signal is therefore around $3 \times 10^8/(3 \times 10^3) = 10^5$ m. Thus, to transmit the human voice directly, the length of the antenna has to be around $\lambda/2 = 50$ km. This is physically impossible. The analog modulation introduced in the later sections of this chapter lifts the frequencies to a higher band so that an antenna with reasonable length can be used.

In the next chapter, we shall introduce digital modulation techniques. The reader will then find that digital modulation is still related to analog modulation, which is why analog modulation techniques are introduced first.

This chapter includes four analog modulation techniques: **amplitude modulation (AM)**, **double-sideband suppressed-carrier (DSB-SC) modulation**, **single-sideband (SSB) modulation**, and **frequency modulation (FM)**.

4.1 Amplitude Modulation

In amplitude modulation, the amplitude of a high-frequency carrier signal is varied in accordance to the amplitude of the modulating message signal. Consider a sinusoidal carrier wave $c(t)$ defined by

$$c(t) = A_c\cos(2\pi f_c t), \tag{4-1}$$

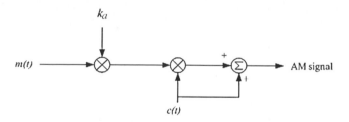

Figure 4-1   Generation of an AM signal

where t represents time in seconds, A_c is the **carrier amplitude** and f_c is the **carrier frequency**. Let $m(t)$ be the modulating message signal (**baseband** signal). The amplitude modulation (AM) signal is expressed as

$$s(t) = A_c[1 + k_a m(t)]\cos(2\pi f_c t), \qquad (4\text{-}2)$$

where k_a is a constant called the **amplitude sensitivity**. Typically, the carrier amplitude A_c and the message signal $m(t)$ are measured in volts and k_a is measured in $volt^{-1}$. Figure 4-1 illustrates a mathematical representation for generating the AM signal.

Figure 4-2(a) shows a baseband signal $m(t)$, and Figures 4-2(b) and 4-2(c) show the corresponding AM wave $s(t)$ for two values of amplitude sensitivity, $k_a = 0.8$ and $k_a = 1.5$. As can be seen in 4-2(b) and 4-2(c), the envelope of $s(t)$ has the same shape as the baseband signal $m(t)$ if $|k_a m(t)| < 1$ for all t.

This condition ensures that the function $1 + k_a m(t)$ is always positive. When the amplitude sensitivity is large enough to make $|k_a m(t)| > 1$ for some t as illustrated in Figure 4-2(c) for t from 0.67 to 0.87 seconds, the carrier wave becomes **overmodulated**, resulting in carrier phase reversals whenever the factor $1 + k_a m(t)$ crosses zero. To avoid overmodulation, the above condition must be satisfied.

The reader should note that the original cosine signal is of the form $c(t) = A_c \cos(2\pi f_c t)$, as indicated in Equation (4-1). The term A_c is the amplitude of $\cos(2\pi f_c t)$ and is a constant. After modulation, the cosine signal becomes

$$s(t) = A_c[1 + k_a m(t)]\cos(2\pi f_c t),$$

as indicated in Equation (4-2). Now, the amplitude of $\cos(2\pi f_c t)$ is no longer a constant but is rather a function of $m(t)$. The term A_c adjusts the magnitude of $m(t)$. Indeed, we may say that the amplitude of $\cos(2\pi f_c t)$ is now of the *form $a + bm(t)$*. We call $A_c[1 + k_a m(t)]$ an *envelope*, which is the dotted curve in Figure 4-2(b).

It is also important that the carrier frequency f_c be greater than the highest frequency component W of the baseband message signal $m(t)$. That is, $f_c > W$. We call W the **message bandwidth**. If this condition is not satisfied, the envelope cannot be visualized and therefore cannot be detected correctly. The reader should remember that the main role of f_c is to lift the baseband signal frequency to a higher one – which will be explained later. Therefore, it has to be higher than W.

The process of demodulation is to recover the original message signal from the modulated signal. An important feature of the AM signal is that the message signal can be easily

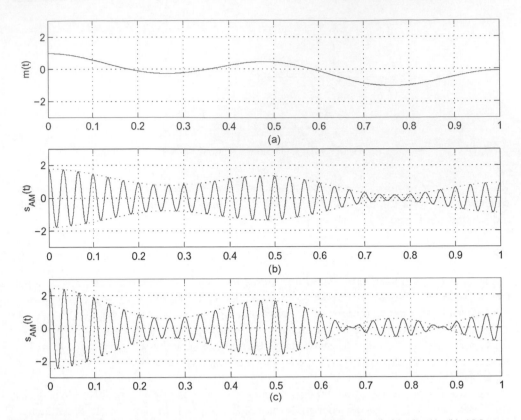

Figure 4-2 Illustration of the amplitude modulation process: (a) baseband signal $m(t)$; (b) AM wave where $|k_a m(t)| < 1$ for all t; (c) AM wave where $|k_a m(t)| > 1$ for some t

recovered by a simple electronic device (a diode) and a parallel connection of an RC (resistor and capacitor) circuit, called *envelope detector*, as shown in Figure 4-3.

In the following, we shall describe the mechanism for an RC circuit to detect the envelope. First, note that an ideal diode has the property that it conducts only in one direction. Let us describe its function in detail.

1. When the input voltage, the voltage of $s(t)$, is larger then the output voltage, the diode will act as a short circuit (it is 'on'). A current will flow from the voltage source to the capacitor as illustrated in Figure 4-4(a). For this period, the capacitor will be charged and the output voltage will be rising. The entire situation is illustrated in region A of Figure 4-4(b).
2. When the input voltage is less then the output voltage, the diode will act as an open circuit (it is 'off'). For this period, there will be no current from the voltage source to the capacitor. But the voltage existing on the capacitor will make the capacitor discharge. That is, although no current can flow through the diode, some current will still flow through the output resister R_L. The output voltage will fall. The entire situation is illustrated in region B of Figure 4-4(b).

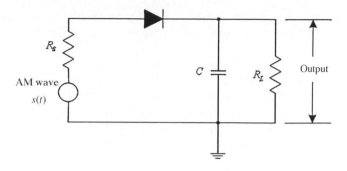

Figure 4-3   An *RC* circuit as an envelope detector

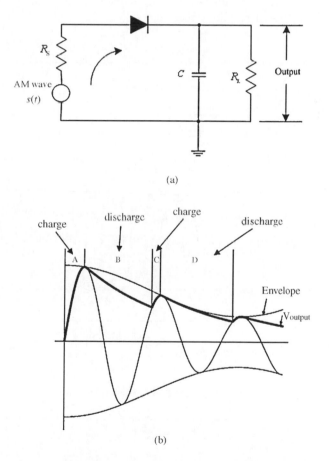

Figure 4-4   The mechanism of the *RC* circuit: (a) the circuit, and (b) the behavior of the circuit in different regions

3. After discharging, the input voltage is larger than the output, the capacitor is then charged and the output voltage will rise again as illustrated in region C of Figure 4-4(b).

In summary, when the input voltage is larger then the output voltage, current flows from the voltage source to the capacitor, the capacitor charges and the output voltage rises. When the input voltage is less then the output voltage, no current flows from the voltage source to the capacitor, the capacitor discharges and the output voltage falls. If we carefully select the values of the resister and capacitor, the rise will not be too slow and the fall will not be too sharp. As illustrated in Figure 4-4(b), the envelope of a modulated signal can now be detected to some degree.

The signals during the demodulation process are illustrated in Figure 4-5: parts (a) and (b) show the message signal and the modulated waveform, respectively. The resulting demodulated signal is shown in 4-5(c). As we discussed before, the charge and discharge speeds, which depend on the values of the resistor and capacitor, should be properly selected, so that the output voltage of this circuit, although a sawtooth one, is a good approximation of the message signal as shown in Figure 4-5(c). This circuit is very simple and therefore is adopted in conventional AM broadcast receivers.

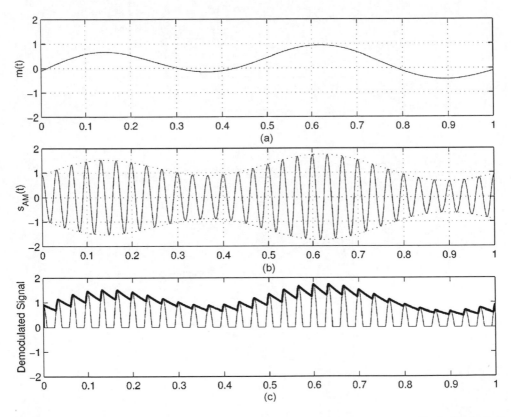

Figure 4-5   The demodulated signal of the envelope detector: (a) the message signal, (b) the AM signal, (c) the demodulated signal

It should be noted that we have introduced only the basic principles of AM demodulation here. In practice it is more complicated, as discussed in Section 4.5.

At the beginning of this chapter, we mentioned that the modulation process lifts the frequency of the baseband to a higher one. This is explained by four different methods as follows.

Method 1

The AM signal $s(t)$ can be expressed as

$$s(t) = A_c\cos(2\pi f_c t) + A_c k_a m(t)\cos(2\pi f_c t). \qquad (4\text{-}3)$$

The Fourier transform of the first term of Equation (4-3), according to Equation (3-145), is

$$\frac{A_c}{2}[\delta(f - f_c) + \delta(f + f_c)].$$

The Fourier transform of the second term, $A_c k_a m(t)\cos(2\pi f_c t)$, can be found by Equation (3-124). Let the Fourier transforms of $x(t)$, $y(t)$ and $z(t)$ be denoted as $X(f)$, $Y(f)$ and $Z(f)$, respectively. Let $z(t) = x(t)y(t)$. Then, according to Equation (3-124):

$$Z(f) = X(f) * Y(f),$$

where $*$ denotes convolution defined in Section 3.4, and

$$X(f) * Y(f) = \int_{-\infty}^{\infty} X(v)Y(f - v)dv.$$

In our case, $x(t) = m(t)$. Thus, $X(f) = M(f)$. We also have $y(t) = \cos(2\pi f_c t)$. Again, according to Equation (3-145):

$$Y(f) = \frac{1}{2}[\delta(f - f_c) + \delta(f + f_c)]$$

$$Y(f - v) = \frac{1}{2}[\delta(f - v - f_c) + \delta(f - v + f_c)].$$

Thus:

$$A_c k_a X(f) * Y(f) = A_c k_a \int_{-\infty}^{\infty} M(v)\frac{1}{2}[\delta(f - v - f_c) + \delta(f - v + f_c)]dv$$

$$= \frac{A_c k_a}{2}[M(f - f_c) + M(f + f_c)]$$

$$\qquad (4\text{-}4)$$

according to Equation (3-137).

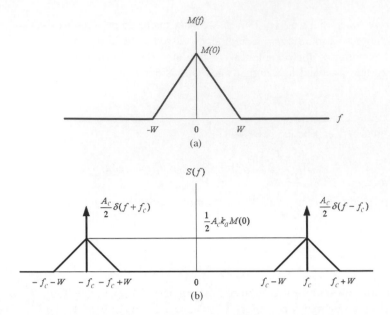

Figure 4-6 The Fourier transform of the AM signals: (a) spectrum of the baseband signal, and (b) spectrum of AM wave

Therefore the Fourier transform of $s(t)$ in Equation (4-3) can be expressed as

$$S(f) = \frac{A_c}{2}[\delta(f - f_c) + \delta(f + f_c)] + \frac{A_c k_a}{2}[M(f - f_c) + M(f + f_c)]. \qquad (4\text{-}5)$$

This means that the baseband signal frequency spectrum $M(f)$ is now moved to $M(f - f_c)$ and $M(f + f_c)$. Assume that the signal $m(t)$ is band-limited to $[-W, W]$; that is, the Fourier transform $M(f)$ is zero for $|f| > W$. The bandwidth of the message signal is then defined as W.

Figure 4-6(a) illustrates the spectrum of the message signal which is bandlimited to $[-W, W]$. For example, a voice signal may have its spectrum concentrated within 3 kHz. The bandwidth of such a voice signal is therefore $W = 3$ kHz. As shown in Figure 4-6(b), the components $M(f - f_c)$ and $M(f + f_c)$ correspond to higher frequencies. The modulated signal then has a spectrum that concentrates around f_c; that is, from $f_c - W$ to $f_c + W$. The bandwidth of the modulated signal is therefore $2W$. Note that lifting a baseband signal frequency to a higher one is our purpose.

Method 2

In Section 3.3, we described the function of convolution in detail. We stated that the convolution mechanism, in certain cases, is like performing a scanning process. Let us consider Figure 4-7. In parts (a) and (b), the Fourier transforms of $m(t)$ and $\cos(2\pi f_c t)$ are shown. As we perform the convolution of these two functions, we may imagine that we first move the two delta functions in Figure 4-7(b) to $-\infty$ and then move these two delta functions from $-\infty$ to the right. As soon as one hits the area where $M(f)$ is not zero, the

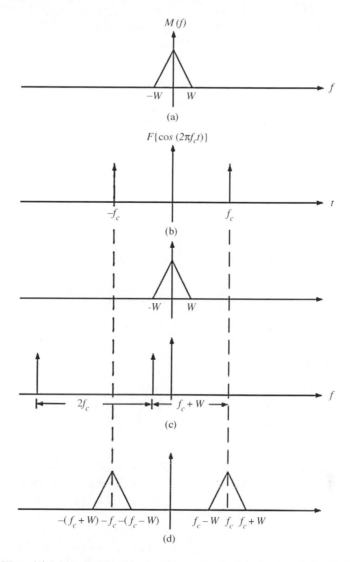

Figure 4-7 AM explained by convolution: (a) the spectrum of the baseband signal; (b) the spectrum of $\cos(2\pi f_c t)$; (c) the sliding of the spectrum of $\cos(2\pi f_c t)$ from left to right; (d) the result of convolution

convolution product starts to have a value – as shown in 4-7(c). The location where the convolution starts to have a nonzero value actually corresponds to how far the delta function is moved to the left. As shown in Figure 4-7(c), the delta function is moved to the left with distance $f_c + W$. Therefore, so far as the convolution is concerned, it starts to have a nonzero value at $f = -(f_c + W)$. Effectively, it moves the entire $M(f)$ to the left to start at $f = -(f_c + W)$ and end at $f = -(f_c - W)$. We may say that $M(f)$ is moved to the left to be centered at $f = -f_c$. It can be easily seen that $M(f)$ will also be moved to the right to be entered at $f = f_c$. This is why the net effect is to move $M(f)$ to the left and to the right, as shown in Figure 4-7(d).

Method 3

Since the main function of amplitude modulation is to multiply $m(t)$ by $\cos(2\pi f_c t)$, let us consider $g(t) = m(t)\cos(2\pi f_c t)$. Since $\cos(\theta) = (1/2)(e^{j\theta} + e^{-j\theta})$ and $F\{e^{j2\pi f_c t}x(t)\} = X(f - f_c)$ according to Equation (3-116), we have

$$G(f) = \frac{1}{2}(M(f - f_c) + M(f + f_c)).   \tag{4-6}$$

Method 4

Let us note that, according to the Fourier transform, the input message signal actually consists of a set of cosine functions. Consider any one of them, denoted as $\cos(2\pi f_m t)$. The modulation process essentially multiplies this signal with $\cos(2\pi f_c t)$:

$$\cos(2\pi f_m t)\cos(2\pi f_c t) = \frac{1}{2}(\cos(2\pi(f_c + f_m)t) + \cos(2\pi(f_c - f_m)t)).   \tag{4-7}$$

Equation (4-7) shows that, after modulation, the original signal becomes two signals, one with frequency $f_c + f_m$ and the other one with frequency $f_c - f_m$. Both frequencies are much higher than f_m. *Thus we conclude that the amplitude modulation lifts all of the frequencies in the original input baseband signal to higher ones*.

The above line of reasoning is a standard trick. To understand the effect of modulation, consider $\cos(2\pi f_m t)$ as the sole input message signal. It will be much easier to understand the effect of modulation this way.

The amplitude modulation therefore translates the frequency spectrum of the message signal $m(t)$ by $+f_c$ and $-f_c$. The modulation essentially shifts the baseband signal from low frequency to high frequency. Consider the human voice. Its frequency may concentrate in the bandwidth from 0 to 5000 Hz. For an AM radio broadcast, the carrier frequency f_c is around 500 kHz to 1.6 MHz, which is much higher.

We want to emphasize again that frequency shifting is very important for wireless communication. As described near the beginning of this chapter, it is not practical to have an antenna that can be used to transmit baseband signals because they have too long wavelengths. To transmit our voice wirelessly, it is absolutely necessary for us to lift the frequency to a higher one so that the wavelength becomes smaller and an antenna can be designed to transmit the higher frequency signals.

Note that the Fourier transform of signals contains negative frequencies. Of course, negative frequencies do not exist physically. They exist in the Fourier transform because we use complex exponentials to represent the signals. Note that a positive frequency corresponds to $e^{j2\pi ft}$. We do need a negative frequency which corresponds to $e^{-j2\pi ft}$. These two terms combine to produce $\cos(2\pi ft)$ as

$$\cos(2\pi ft) = \frac{1}{2}(e^{j2\pi ft} + e^{-j2\pi ft}).$$

Finally, we want to emphasize again the meanings of the bandwidth of the message signal, the carrier frequency, and the bandwidth of the transmitted signal. Figure 4-6 shows the corresponding input and output spectra for a modulator.

- *Bandwidth of the baseband message signal.* The bandwidth of the message signal is defined as the bandwidth of the baseband message signal, denoted as $m(t)$, to be transmitted. For example, the message signal $m(t)$ for AM radio broadcasting is limited to a bandwidth of approximately 5000 Hz, i.e. with $W = 5\,\text{kHz}$.
- *Location of the carrier frequency.* The location of the carrier frequency, denoted as f_c, is the target frequency that the modulators want to lift to. For example, an AM broadcasting system utilizes the frequency band 535–1605 kHz for transmission of voice and music. The carrier frequency f_c is allocated from 540 kHz to 1600 kHz with 10 kHz spacing.
- *Bandwidth of the transmitted signal.* After modulation, the modulated signal has spectrum concentrated over the carrier frequency f_c. The bandwidth of the modulated signal is defined as the bandwidth for which most of the signal concentrates on. For example, for conventional AM, as shown in Figure 4-6(b), the bandwidth of the transmitted signal is $2W$. Therefore, for AM broadcasting, the bandwidth of the transmitted signal is $2 \times 5\,\text{kHz} = 10\,\text{kHz}$.

What is the significance of the bandwidth of the transmitted signal? If the bandwidth is wide, this means that the transmitting devices must be able to cope with a wide band of signals. If the devices fail to do so, some signals corresponding to certain frequencies will be distorted. For example, one can easily understand that some type of amplifier must be used. If the gain of the amplifier is not uniform for the frequencies within the bandwidth of the transmitted signal, this is far from ideal. The reader can probably tolerate human voices being distorted a little, but not pictures or music. The situation is the same at the receiving end. If the receiving antenna, for example, responds poorly to some frequencies, the received signal will be distorted.

Perhaps the concept of bandwidth can be understood by thinking about your speakers. If you want to listen to good-quality music, you need good speakers. If a speaker is good, it can handle signals with a wide range of frequencies. Thus a good speaker has a wide bandwidth. On the other hand, if your speaker has only a narrow bandwidth, it cannot handle a large range of frequencies, so you cannot listen to good music.

4.2 Double-sideband Suppressed Carrier (DSB-SC)

This method is similar to the AM technique, except that the constant carrier term is suppressed. The name 'double sideband' will be explained when we introduce the spectrum of a DSB-SC signal.

As will be shown in this section, it is possible to recover the message signal without the constant carrier, so it is possible to save transmission power by suppressing the constant carrier. The DSB-SC signal is expressed as

$$s(t) = A_c m(t)\cos(2\pi f_c t). \tag{4-8}$$

Thus, this technique translates the frequency spectrum of the message signal $m(t)$ by multiplying it by a sinusoidal signal of frequency f_c. The modulated signal $s(t)$ undergoes a

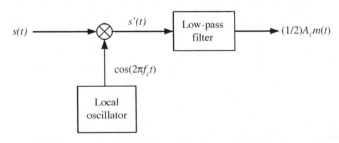

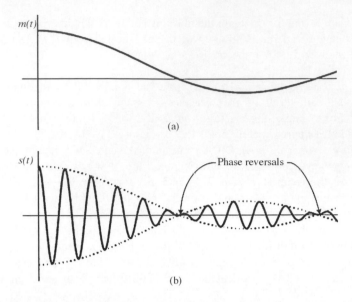

Figure 4-8 Double-sideband modulation: (a) the message signal, and (b) the DSB-SC modulated signal

phase reversal whenever the message signal $m(t)$ crosses zero, as illustrated in Figure 4-8. The multiplication of $\cos(2\pi f_c t)$ by $m(t)$ is equivalent to varying the carrier amplitude in proportion to $m(t)$.

Figure 4-8 shows that the simple diode-circuit demodulation technique cannot be used in this case because we cannot detect the message signal correctly by using an envelope detector. Some other method must be used for demodulation.

The system required to recover the signal $m(t)$ from the DSB-SC signal, the demodulator, is shown in Figure 4-9. The local oscillator produces a sinusoidal waveform $\cos(2\pi f_c t)$ with the same frequency and phase as the transmitted carrier. This method of recovering the original signal requires an exact match of the carrier frequency and phase with the transmitting end. Therefore, this method is called *synchronous demodulation* or *coherent demodulation*. Why this system works is explained in the following.

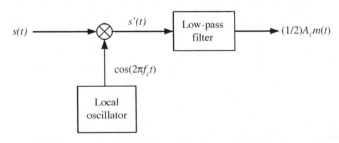

Figure 4-9 The demodulator: coherent detection of the DSB-SC modulated signal

The signal $s'(t) = s(t)\cos(2\pi f_c t)$ indicated in Figure 4-9 can be expressed as

$$
\begin{aligned}
s'(t) &= s(t)\cos(2\pi f_c t)\\
&= A_c m(t)\cos^2(2\pi f_c t)\\
&= A_c m(t)\frac{1 + \cos(2\pi(2f_c)t)}{2} \qquad\qquad (4\text{-}9)\\
&= \frac{A_c}{2}m(t) + \frac{A_c}{2}m(t)\cos(2\pi(2f_c)t).
\end{aligned}
$$

The first term is a low-frequency component which is proportional to $m(t)$. The second term is a high-frequency component which is equivalent to a DSB-SC signal with carrier frequency $2f_c$. The signal $s'(t)$ is then passed through a low-pass filter. This will pass only the low-frequency component that corresponds to $m(t)$ and reject the high-frequency component in the DSB-SC signal. Therefore, the demodulated output of the low-pass filter contains only the term $(A_c/2)m(t)$. This means that $m(t)$ is recovered.

The Fourier transform of $s(t)$ can be expressed as

$$
S(f) = \frac{A_c}{2}[M(f - f_c) + M(f + f_c)]. \qquad\qquad (4\text{-}10)
$$

The DSB-SC modulation therefore translates the frequency spectrum by $\pm f_c$, as illustrated in Figure 4-10. As compared with Figure 4-6, the DSB-SC signal has no constant carrier which

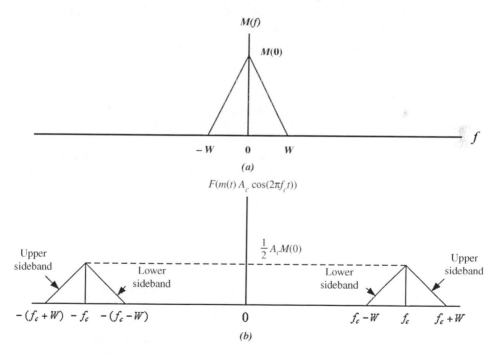

Figure 4-10   The spectra of DSB signals: (a) spectrum of the message signal, and (b) spectrum of the DSB-SC modulated signal

corresponds to delta functions in Figure 4-6. In addition, the spectrum of the DSB-SC signal is divided into two parts: the lower sideband and the upper sideband. As shown in Figure 4-10, the lower sideband of $S(f)$ corresponds to the part of $S(f)$ with $|f| < f_c$ and the upper sideband corresponding to the part of $S(f)$ with $|f| > f_c$. Therefore, this modulation technique is termed **double-sideband modulation**. In the next section, we will introduce another modulation technique, termed 'single-sideband modulation' (SSB), which transmits only the upper sideband or lower sideband of $S(f)$.

We have introduced the concept of the demodulation of a DSB-SC signal based on the analysis of time-domain signals. However, for a better understanding of the demodulation process, we will use the Fourier transform to explain the demodulation process of a DSB-SC signal.

From the spectrum shown in Figure 4-10, to recover the original signal from $s(t)$ it is necessary to retranslate the spectrum to its original position. The process is referred to as *demodulation* or *detection*.

The spectrum of the modulated waveform can be retranslated to the original position by multiplying the modulated signal by $\cos(2\pi f_c t)$ at the receiving end, as shown in Figure 4-9. The resulting signal was denoted as $s'(t)$ in Equation (4-9). This process can be considered as a demodulation of the signal which will cause spectrum translation.

By examining Equation (4-9), we can easily see that the Fourier transform of $s'(t)$ is

$$S'(f) = \frac{A_c}{2} M(f) + \frac{A_c}{4} [M(2f - f_c) + M(2f_c + f)]. \qquad (4\text{-}11)$$

A little reflection shows that $s'(t)$ yields the spectrum shown in Figure 4-11. Now the original signal $M(f)$ is located in the low-frequency part. Therefore, the message signal can be recovered by using a low-pass filter which allows $M(f)$ to pass and removes the components centered around $\pm 2f_c$.

Let us summarize the principle of this kind of modulation and demodulation as follows:

- *Given a message signal $m(t)$, the modulation process multiplies $m(t)$ by $\cos(2\pi f_c t)$. This lifts the frequency of the spectrum $M(f)$ of the message signal to $M(f - f_c)$ and $M(f + f_c)$.*
- *After the receiver receives the signal, the demodulation process multiplies the received signal by $\cos(2\pi f_c t)$ again. This transforms the spectrum of the received signal, say*

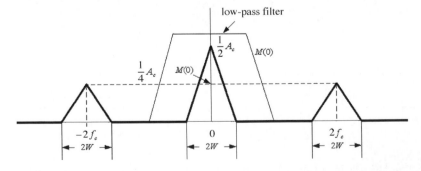

Figure 4-11   Illustration of the spectrum of the demodulation process of a DSB signal

$M(f_c + f)$, *to $M(f)$ and $M(f + 2f_c)$, and also transforms the spectrum of the received signal, say $M(f_c - f)$, to $M(f)$ and $M(f - 2f_c)$. The original signal $M(f)$ appears in the low-frequency part. This means that the original signal $M(f)$ can be recovered by a low-pass filter.*

The reader should remember that both modulation and demodulation make use of multiplying a signal by $\cos(2\pi f_c t)$. For modulation, this multiplication lifts the frequency to a higer one and the demodulation recovers it.

Just as with amplitude modulation, double-sideband modulation produces a signal centered at the carrier frequency f_c with a bandwidth $2W$, where W is the bandwidth of the baseband signal.

4.3 Single-sideband (SSB) Modulation

To further reduce the required transmission power and bandwidth of the modulated signal, it is possible to transmit only the upper sideband or lower sideband of the DSB-SC signal, *because both sidebands carry the same information*. This technique is called *single-sideband modulation*.

Let us start with considering a much simplified form of input. That is, imagine that our input message signal is $\cos(2\pi f_m t)$. After DSB-SC modulation, the modulated signal becomes

$$s(t) = \cos(2\pi f_m t)\cos(2\pi f_c t). \tag{4-12}$$

But

$$s(t) = \cos(2\pi f_m t)\cos(2\pi f_c t) - \frac{1}{2}(\cos(2\pi(f_c + f_m)t) + \cos(2\pi(f_c - f_m)t)). \tag{4-13}$$

The demodulation process multiplies $s(t)$ by $\cos(2\pi f_c t)$. Thus, we have

$$
\begin{aligned}
s'(t) &= s(t)\cos(2\pi f_c t) \\
&= \frac{1}{2}(\cos(2\pi(f_c + f_m)t)\cos(2\pi f_c t) + \cos(2\pi(f_c - f_m)t)\cos(2\pi f_c t)) \\
&= \frac{1}{4}(\cos(2\pi(2f_c + f_m)t) + \cos(2\pi f_m t) + \cos(2\pi(2f_c - f_m)t) + \cos(2\pi f_m t)) \\
&= \frac{1}{2}\cos(2\pi f_m t) + \frac{1}{4}(\cos(2\pi(2f_c + f_m)t) + \cos(2\pi(2f_c - f_m)t)).
\end{aligned}
\tag{4-14}
$$

We have recovered $\cos(2\pi f_m t)$ successfully. The question is: Do we really need both $\cos(2\pi(f_c + f_m)t)$ and $\cos(2\pi(f_c - f_m)t)$ to recover $\cos(2\pi f_m t)$? The answer is no. In the following, we shall show that we need either of them to recover $\cos(2\pi f_m t)$.

Suppose that we use only $\cos(2\pi(f_c + f_m)t)$ to recover the original signal. Then:

$$s(t) = \cos(2\pi(f_c + f_m)t) \tag{4-15}$$
$$s'(t) = \cos(2\pi(f_c + f_m)t)\cos(2\pi f_c t)$$
$$= \frac{1}{2}(\cos(2\pi(2f_c + f_m)t) + \cos(2\pi f_m t)). \tag{4-16}$$

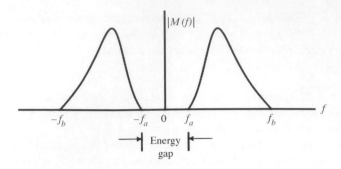

Figure 4-12   Spectrum of a message signal $m(t)$ with an energy gap centered around the origin

As one can see, $\cos(2\pi f_m t)$ is now recovered successfully. Using a similar derivation, we can show that $\cos(2\pi f_m t)$ can be recovered by using $\cos(2\pi(f_c - f_m)t)$ only.

The required transmission bandwidth of a DSB-SC signal is $2W$, where W is the message bandwidth; while for the SSB signal the required transmission bandwidth is only W.

For generation of an SSB modulated signal to be possible, the message spectrum should have an energy gap centered at the origin, as illustrated in Figure 4-12. This constraint is naturally satisfied by voice signals, whose energy gap is about 600 Hz (i.e. from -300 to $+300$ Hz).

4.3.1 Generating the SSB Signal

We next introduce two methods to generate the SSB signal.

Method 1: The Bandpass Filter

The message signal is first modulated using the DSB-SC technique. By passing the DSB-SC signal through a bandpass filter, we may generate an SSB signal containing the upper sideband of the DSB-SC signal. Generation of the SSB signal is shown in Figure 4-13. Assume that the spectrum of the message signal is as shown in Figure 4-14(a). The resulting spectrum of the DSB-SC signal and the frequency response of the required bandpass filter are shown in Figure 4-14(b). The filter will pass the frequency components of the DSB-SC signal within $f_c + f_a$ and $f_c + f_b$ and reject all the frequency components out of this range.

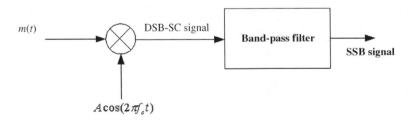

Figure 4-13   Generation of an SSB signal by the band-pass filter method

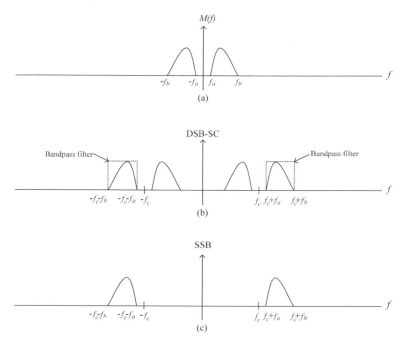

Figure 4-14 Generation of an SSB signal by the band-pass filter method: (a) message spectrum; (b) spectrum of the DSB-SC signal and the idealized frequency response of the required band-pass filter; (c) spectrum of an SSB signal containing the upper sideband

Thus given the message spectrum defined in Figure 4-14(a), we find that the corresponding spectrum of the SSB signal is as shown in Figure 4-14(c).

Demodulation of an SSB signal is exactly the same as for the DSB-SC signal. Recovery of the message signal is illustrated in Figure 4-15. The spectrum of the message signal is shown in 4-15(a). Let $s(t)$ be the resulting SSB signal; the spectrum of this is shown in 4-15(b). The spectrum of the product $s(t) \times \cos(2\pi f_c t)$ is obtained by shifting $(1/2)S(f)$ to the right by f_c and shifting $(1/2)S(f)$ to the left by f_c, as for DSB-SC modulation. The resulting spectrum of $s(t) \times \cos(2\pi f_c t)$ is shown in 4-15(c). It is important to note that the original spectrum of the message signal appears in the low-frequency part. Therefore, the message can be recovered by passing the signal through a low-pass filter, and the demodulated signal is illustrated in 4-15(d). Although the receiver receives half of the frequency components, the demodulator successfully recovers the message signal $m(t)$.

In the following we introduce another method to produce an SSB signal.

Method 2: The Fourier Transform

Before introducing this method formally, let us start with one simple case. Let us suppose that our input message is only $\cos(2\pi f_m t)$. That is:

$$m(t) = \cos(2\pi f_m t). \tag{4-17}$$

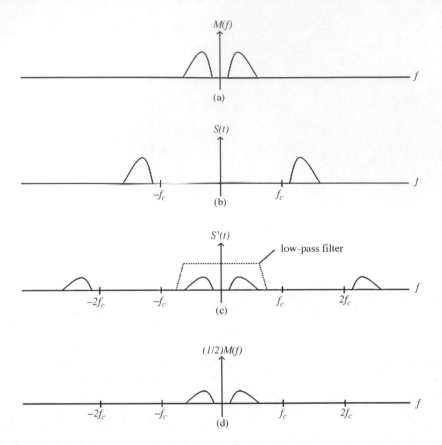

Figure 4-15 Demodulation of an SSB signal: (a) spectrum of the message signal $m(t)$; (b) spectrum of the SSB signal containing the upper sideband; (c) spectrum of $s(t) \times \cos(2\pi f_c t)$; (d) the resulting demodulated signal

To send this message signal we do not modulate it directly by multiplying it by $\cos(2\pi f_c t)$. Instead, we first perform a $\pi/2$ phase shift on $m(t)$, so that it becomes $\sin(2\pi f_m t)$. Denote this by $\hat{m}(t)$:

$$\hat{m}(t) = \sin(2\pi f_m t). \tag{4-18}$$

The modulated signal will be as follows:

$$\begin{aligned} s(t) &= m(t)\cos(2\pi f_c t) \mp \hat{m}(t)\sin(2\pi f_c t) \\ &= \cos(2\pi f_m t)\cos(2\pi f_c t) \mp \sin(2\pi f_m t)\sin(2\pi f_c t) \\ &= \cos(2\pi(f_c \pm f_m)t). \end{aligned} \tag{4-19}$$

The sign $\mp$ means that either minus or plus is used. If minus is used, we send $\cos(2\pi(f_c + f_m)t)$, and if plus is used we send $\cos(2\pi(f_c - f_m)t)$. In other words, we send signals only in one sideband.

We still face one problem. In general, $m(t)$ consists of a lot of cosine functions. How can we perform a $\pi/2$ phase shift on every cosine function? Essentially, we do this by performing a special kind of transform. The Fourier transform of $m(t) = \cos(2\pi f_m t)$ is $M(f) = (1/2)(\delta(f - f_m) + \delta(f + f_m))$. Suppose we multiply the spectrum of the message signal $M(f)$ by $-j$ for $f > 0$ and multiply by $+j$ for $f < 0$. This involves multiplying $M(f)$ by the following function:

$$H(f) = \begin{cases} -j & \text{for } f > 0 \\ j & \text{for } f < 0. \end{cases} \tag{4-20}$$

Thus we have

$$\hat{M}(f) = M(f)H(f) = \frac{1}{2j}(\delta(f - f_m) - \delta(f + f_m)). \tag{4-21}$$

The inverse Fourier transform of $\hat{M}(f)$ is therefore $\hat{m}(t) = \sin(2\pi f_m t)$ according to Equation (3-146).

Let us summarize the above discussion as follows: Given $m(t)$, $\hat{m}(t)$ is obtained from $m(t)$ by phase-shifting by $\pi/2$ every cosine function that it contains. The signal $\hat{m}(t)$ can be obtained by the following steps:

1. Find the Fourier transform $M(f)$ of $m(t)$.
2. For every $f > 0$, multiply $M(f)$ by $-j$, and for every $f < 0$ multiply $M(f)$ by $+j$. The result is $\hat{M}(f)$.
3. Find the inverse Fourier transform of $\hat{M}(f)$. The result is $\hat{m}(t)$.

In practice we cannot find the Fourier transform of a signal. Therefore, in step 1 we may perform a discrete Fourier transform. In step 2, for every $i < (n/2) - 1$ multiply A_i by $-j$, and for every $i > (n/2) - 1$ multiply A_i by $+j$, where A_i is obtained in the discrete Fourier transform. We also perform an inverse discrete Fourier transform in step 3.

An SSB signal can be mathematically expressed as

$$s(t) = A_c[m(t)\cos(2\pi f_c t) \mp \hat{m}(t)\sin(2\pi f_c t)], \tag{4-22}$$

where the minus sign is used for upper-sideband SSB and the plus sign is used for lower-sideband SSB. To produce the signal $\hat{m}(t)$, we can perform $\hat{M}(f) = M(f)H(f)$ in the frequency domain. Let $h(t)$ be the inverse Fourier transform of $H(f)$. It can be shown that $h(t) = 1/(\pi t)$ for $H(f)$ given in Equation (4-20). The term $\hat{m}(t)$ can be obtained in the time domain by convoluting $m(t)$ with $h(t)$:

$$\begin{aligned} \hat{m}(t) &= m(t) * h(t) \\ &= \int_{-\infty}^{\infty} \frac{1}{\pi(\tau - t)} m(\tau)d\tau. \end{aligned} \tag{4-23}$$

In summary, the Fourier transform of $\hat{m}(t)$, according to Equation (3-121), can be expressed as

$$\hat{M}(f) = M(f)H(f), \tag{4-24}$$

where

$$H(f) = \begin{cases} -j & \text{for } f > 0 \\ j & \text{for } f < 0. \end{cases} \qquad (4\text{-}25)$$

To prove that $h(t) = 1/(\pi t)$ is the inverse Fourier transform of the function expressed in Equation (4-25) is beyond the scope of this book. In the following we shall explain why Equation (4-22) works; that is, why $s(t)$ in (4-22) has only one half of the frequency components of those of Equation (4-10) expressing the DSB signal. This is done through Fourier transform analysis of the formula in Equation (4-22).

The Fourier transform of the first term in Equation (4-22) is illustrated in Figure 4-16(b), of which, for simplicity, $M(f)$ is assumed to be a real positive function as shown in 4-16(a).

The Fourier transform of the second term in Equation (4-22) is a little bit harder to obtain. Let us rewrite it as

$$\hat{m}(t)\sin(2\pi f_c t).$$

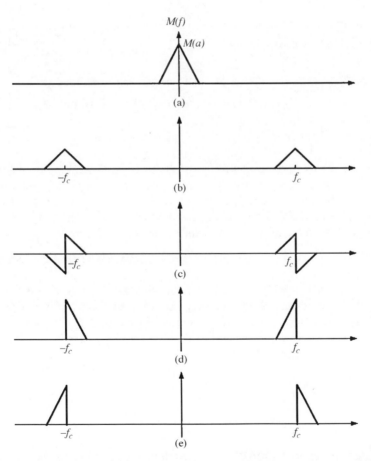

Figure 4-16 Generation of an SSB signal by the Fourier transform method: (a) spectrum of the message signal $m(t)$; (b) spectrum of the DSB signal; (c) spectrum of $z(t)$; (d) spectrum of the lower-sideband SSB; (e) spectrum of the upper-sideband SSB

The above expression is a product of two functions of time. The Fourier transform of it can be obtained as follows. Let

$$z(t) = \hat{m}(t)\sin(2\pi f_c t).$$ (4-26)

According to Equations (3-124) and (3-146), we have

$$Z(f) = \hat{M}(f) * \frac{1}{2j}[\delta(f - f_c) - \delta(f + f_c)],$$ (4-27)

where * denotes convolution, which was defined in Section 3.6. It can be easily proved that

$$Z(f) = \frac{1}{2j}(\hat{M}(f - f_c) - \hat{M}(f + f_c)).$$ (4-28)

What is $\hat{M}(f)$? The term can be found in Equations (4-24) and (4-25). Consider the case where $f_c \gg W$, where W is the bandwidth of the baseband signal. For the case $f > 0$, only $\hat{M}(f - f_c)$ needs to be considered because $\hat{M}(f + f_c)$ will not interfere with $\hat{M}(f - f_c)$. Let us consider the following two cases separately.

Case 1: $f > f_c$. In this case, according to Equations (4-24) and (4-25):

$$\hat{M}(f - f_c) = -jM(f - f_c)$$
$$\frac{1}{2j}\hat{M}(f - f_c) = -\frac{1}{2}M(f - f_c).$$ (4-29)

Case 2: $0 < f < f_c$. In this case, according to Equations (4-24) and (4-25), we can prove that

$$\frac{1}{2j}\hat{M}(f - f_c) = \frac{1}{2j}jM(f - f_c) = \frac{1}{2}M(f - f_c).$$ (4-30)

Similarly, for $-f_c < f < 0$ we have $Z(f) = (1/2)M(f + f_c)$, and for $f < -f_c$ we have $Z(f) = -(1/2)M(f + f_c)$. In summary:

$$Z(f) = \begin{cases} -(1/2)M(f - f_c), & f_c < f \\ +(1/2)M(f - f_c), & 0 < f < f_c \\ +(1/2)M(f + f_c), & -f_c < f < 0 \\ -(1/2)M(f + f_c), & f < -f_c \end{cases}$$ (4-31)

Therefore, we can illustrate the spectrum $Z(f)$ as in Figure 4-16(c). Let us rewrite Equation (4-22) again as follows:

$$s(t) = A_c[m(t)\cos(2\pi f_c t) \mp \hat{m}(t)\sin(2\pi f_c t)].$$ (4-32)

Note that the Fourier transform of the first term is illustrated in Figure 4-16(b) which corresponds to the spectrum of a DSB signal. Combining Figures 4-16(b) and (c), we can see that one sideband will be canceled out no matter whether the sign of the second term is positive or negative. If the sign is positive, it corresponds to lower-sideband SSB as shown in 4-16(d). If the sign is negative, it corresponds to upper-sideband SSB as shown in 4-16(e).

4.3.2 The Balanced Modulator

A technique for generating an SSB signal is the balanced modulator shown in Figure 4-17. This modulator is a direct implementation of Equation (4-22). The modulating signal is split into two identical signals. One modulates the $\cos(2\pi f_c t)$ and the other passes through a $-90°$ phase shift before modulating the $\sin(2\pi f_c t)$ carrier. The sign used for the $\sin(2\pi f_c t)$ carrier determines whether an upper SSB or lower SSB is transmitted. It should be emphasized here that the 90° shift can be done by the steps presented above.

4.3.3 Demodulation

The demodulation of Method 2 is the same as for Method 1 – namely, multiplying the received signal by $\cos(2\pi f_c t)$. Note that

$$s(t) = m(t)\cos(2\pi f_c t) + \hat{m}(t)\sin(2\pi f_c t). \tag{4-36}$$

Thus, multiplying the received signal by $\cos(2\pi f_c t)$, we have

$$
\begin{aligned}
s'(t) &= s(t)\cos(2\pi f_c t) \\
&= m(t)\cos^2(2\pi f_c t) + \hat{m}(t)\sin(2\pi f_c t)\cos(2\pi f_c t) \\
&= m(t)\frac{1}{2}(1 + \cos(2\pi(2f_c)t)) + \hat{m}(t)\frac{1}{2}\sin(2\pi(2f_c)t) \\
&= \frac{1}{2}m(t) + \frac{1}{2}\underbrace{(m(t)\cos(2\pi(2f_c)t) + \hat{m}(t)\sin(2\pi(2f_c)t))}_{\text{SSB high frequency signal}}.
\end{aligned}
\tag{4-37}
$$

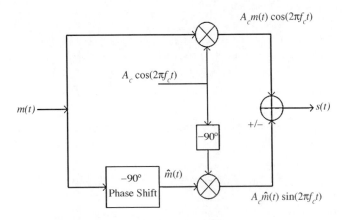

Figure 4-17   An SSB balanced modulator

Equation (4-37) shows that the signal contains a message $m(t)$ and another SSB signal with carrier frequency $2f_c$. Since $f_c \gg W$, the message signal can be recovered by a low-pass filter as pointed out before.

4.3.4 Bandwidth of the SSB Signal

Finally, we can see that single-sideband modulation produces a transmitted signal with a bandwidth W. The frequencies of the transmitted signals are either from $f_c - W$ to f_c or from f_c to $f_c + W$.

4.4 Frequency Modulation (FM)

In earlier sections we have investigated the effect of modulating the *amplitude* of a sinusoidal carrier wave by the message signal. There is another way of modulating a sinusoidal carrier wave, namely *frequency modulation*, in which the frequency of the carrier wave varies according to the message signal. The amplitude of the modulated carrier is now maintained constant. Frequency modulation can provide better communication quality then amplitude modulation because there is less noise after demodulation. *Roughly speaking, when the received signals of FM and AM systems are corrupted by noise of the same level, the amplitude of the carrier is more sensitive than the frequency of the carrier to noise corruption. Therefore, in the demodulation of an FM signal, the noise output is less than that of an AM, DSB-SC or SSB signal. This improvement in performance is achieved at the expense of increased transmission bandwidth.*

Let $\theta_i(t)$ denote the angle of a modulated signal:

$$s(t) = A_c \cos[\theta_i(t)], \tag{4-38}$$

where A_c is the carrier amplitude and $\theta_i(t)$ varies according to the message signal $m(t)$. To appreciate the concept of frequency variation, it is necessary to define the *instantaneous frequency*. For an ordinary fixed-frequency sinusoidal signal, we have $\theta_i(t) = 2\pi f_c t$. The instantaneous frequency is f_c for this signal, since the frequency remains constant for all t. But it is possible that the frequency changes with time. Under those circumstances, we may define the *average frequency* over an interval from t to $t + \Delta t$ as

$$f_{\Delta t}(t) = \frac{\theta_i(t + \Delta t) - \theta_i(t)}{2\pi \Delta t}. \tag{4-39}$$

The instantaneous frequency of the signal $f_i(t)$ is defined as

$$f_i(t) = \lim_{\Delta t \to 0} f_{\Delta t}(t) = \frac{1}{2\pi} \frac{d\theta_i(t)}{dt}. \tag{4-40}$$

The frequency-modulated signal is then formed by varying the instantaneous frequency directly with the modulating signal given by

$$f_i(t) = f_c + k_f m(t). \tag{4-41}$$

The term f_c represents the frequency of the unmodulated carrier, and the constant k_f represents the **frequency sensitivity** of the modulator, expressed in hertz per volt under the assumption that $m(t)$ is a voltage waveform. The maximum of the term $|k_f m(t)|$ in Equation (4-41) is called frequency deviation. One can easily see that $|k_f m(t)|$ has to be limited in practice.

Substituting Equation (4-41) into (4-40) and integrating both sides with respect to time, we get

$$\theta_i(t) = 2\pi f_c t + 2\pi k_f \int_{-\infty}^{t} m(\tau)d\tau. \tag{4-42}$$

The frequency-modulated signal is therefore described in the time domain by substituting Equation (4-42) into (4-38) as follows:

$$s(t) = A_c \cos[2\pi f_c t + 2\pi k_f \int_{-\infty}^{t} m(\tau)d\tau]. \tag{4-43}$$

Figure 4-18 shows the relationship between the message signal and modulated FM waveform. Figure 4-18(a) represents a message signal $m(t)$ which is assumed to be a low-frequency sinusoid. Figure 4-18(b) illustrates the carrier waveform in which the zero crossing points have a perfect regularity in their spacing. The resulting FM signal is shown in 4-18(c): the zero crossings no longer have regular spacing, a consequence of $\theta_i(t)$ varying with $m(t)$. Another important feature is that the *envelope* of an FM signal is constant, whereas the envelope of an AM, DSB-SC or SSB signal depends on the message signal.

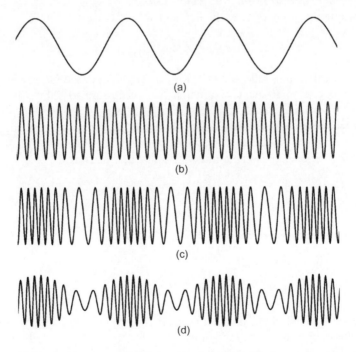

Figure 4-18 An FM signal: (a) a sinusoidal message signal; (b) the carrier wave; (c) frequency-modulated signal; (d) FM signal after differentiator

Since the FM signal $s(t)$ defined in Equation (4-43) is a nonlinear function of the modulating signal $m(t)$, the spectrum of an FM signal is not related in a simple manner to the message signal. Therefore, spectrum analysis of an FM signal is much more difficult than for AM and DSB-SC signals. Here, this problem was solved by assuming that the message signal is sinusoidal. *The result shows that the bandwidth of the spectrum of a sinusoidally modulated FM signal is infinity, so that the bandwidth required to transmit such a signal is similarly infinity. However, in practice, the FM signal can be limited to a finite bandwidth if we allow a specified amount of distortion.*

The following text related to the bandwidth with frequency modulation is quite difficult. Assume that the message signal is $m(t) = A_m\cos(2\pi f_m t)$. From Equation (4-42) we have

$$\theta_i(t) = 2\pi f_c t + \frac{k_f A_m}{f_m}\sin(2\pi f_m t). \tag{4-44}$$

Define $\Delta f = k_f A_m$. The quantity Δf is called the *frequency deviation* which represents the maximum deviation of the instantaneous frequency from the carrier frequency f_c. The frequency deviation is proportional to the amplitude of the modulating signal. The ratio of the frequency deviation to the modulation frequency f_m, denoted as β, is called the *modulation index* of the FM signal. The modulation index is defined by

$$\beta = \frac{\Delta f}{f_m}. \tag{4-45}$$

Therefore we have

$$\theta_i(t) = 2\pi f_c t + \beta\sin(2\pi f_m t). \tag{4-46}$$

The FM signal can then be written as

$$s(t) = A_c\cos(2\pi f_c t + \beta\sin(2\pi f_m t)). \tag{4-47}$$

We may simplify the analysis by using the complex representation of band-pass signals; that is:

$$\begin{aligned}s(t) &= \mathrm{Re}[A_c\exp(j2\pi f_c t + j\beta\sin(2\pi f_m t))] \\ &= \mathrm{Re}[\tilde{s}(t)\exp(j2\pi f_c t)]\end{aligned} \tag{4-48}$$

where $\tilde{s}(t)$ is given by

$$\tilde{s}(t) = A_c\exp[j\beta\sin(2\pi f_m t)]. \tag{4-49}$$

Since $\sin(2\pi f_m t)$ is periodic with period $T_m = 1/f_m$, the signal $\tilde{s}(t)$ is a periodic function of time with a fundamental frequency equal to f_m. Therefore, we may expand $\tilde{s}(t)$ in the form of a complex Fourier series:

$$\tilde{s}(t) = \sum_{n=-\infty}^{\infty} c_n\exp(j2\pi nf_m t), \tag{4-50}$$

where the complex Fourier coefficients c_n are as follows:

$$
c_n = f_m \int_0^{1/f_m} \tilde{s}(t) \exp(-j2\pi n f_m t) dt
$$

$$
= f_m A_c \int_0^{1/f_m} \exp(j\beta \sin(2\pi f_m t) - j2\pi n f_m t) dt. \tag{4-52}
$$

Define a new variable $x = 2\pi f_m t$. Hence we may rewrite c_n in the form

$$
c_n = \frac{A_c}{2\pi} \int_0^{2\pi} \exp[j(\beta \sin x - nx)] dx. \tag{4-53}
$$

Define the function $J_n(\beta)$ as

$$
J_n(\beta) = \frac{1}{2\pi} \int_0^{2\pi} \exp[j(\beta \sin x - nx)] dx. \tag{4-54}
$$

The function $J_n(\beta)$ is known as the nth order Bessel function of the first kind. Accordingly, we may express the Fourier coefficients as

$$
c_n = A_c J_n(\beta). \tag{4-55}
$$

Equation (4-50) can be now expressed as

$$
\tilde{s}(t) = A_c \sum_{n=-\infty}^{\infty} J_n(\beta) \exp(j2\pi n f_m t). \tag{4-56}
$$

Now, substituting Equation (4-56) into (4-48), we have

$$
s(t) = A_c \mathrm{Re}\left[\sum_{n=-\infty}^{\infty} J_n(\beta) \exp[j2\pi(f_c + n f_m)t] \right]
$$

$$
= A_c \sum_{n=-\infty}^{\infty} J_n(\beta) \cos[2\pi(f_c + n f_m)t]. \tag{4-57}
$$

Equation (4-57) is the desired form for the Fourier series representation of the single-tone modulated FM signal. The spectrum of $s(t)$ can be obtained by taking the Fourier transform of it. We thus have

$$
S(f) = \frac{A_c}{2} \sum_{n=-\infty}^{\infty} J_n(\beta)[\delta(f - f_c - n f_m) + \delta(f + f_c + n f_m)]. \tag{4-58}
$$

Figure 4-19 plots the Bessel function $J_n(\beta)$ versus the modulation index β for different positive integer values of n. For negative n, we have

$$
J_{-n}(\beta) = \begin{cases} J_n(\beta) & n \text{ even} \\ -J_n(\beta) & n \text{ odd}. \end{cases} \tag{4-59}
$$

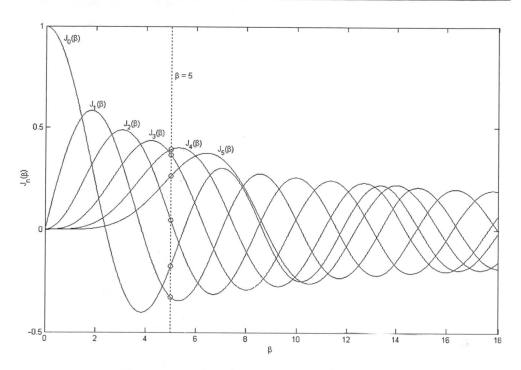

Figure 4-19 Plots of Bessel functions of the first kind

The result shows that the spectrum of an FM signal contains carrier components located at f_c and $-f_c$ and an infinite set of side frequencies located symmetrically around f_c and $-f_c$ with frequency separations of $\pm f_m, \pm 2f_m, \cdots$. In addition, the amplitude of the frequency components varies with β according to $J_n(\beta)$.

For example, let the message signal be $m(t) = \cos(20\pi t)$. Further assume that the message signal is used to frequency-modulate the carrier with $k_f = 50$. The modulation index is then given by $\beta = k_f/f_m = 50/10 = 5$. Let $A_c = 1$. The resulting FM signal is

$$s(t) = \cos(2\pi f_c t + 5\sin(20\pi t)). \tag{4-60}$$

The FM signal can also be expressed as

$$s(t) = \sum_{n=-\infty}^{\infty} J_n(5)\cos(2\pi(f_c + 10n)t). \tag{4-61}$$

The Fourier transform of $s(t)$ is then

$$S(f) = \sum_{n=-\infty}^{\infty} J_n(5)[\delta(f - f_c - 10n) + \delta(f + f_c + 10n)]. \tag{4-62}$$

The first few values of $J_n(5)$ can be obtained from Figure 4-19 as the crossing points between the dashed line at $\beta = 5$ with the function curves and are denoted by circles in the figure. Figure 4-20 shows the amplitude spectrum $|S(f)|$ of $S(f)$ with $f_c = 300\,\text{Hz}$. The

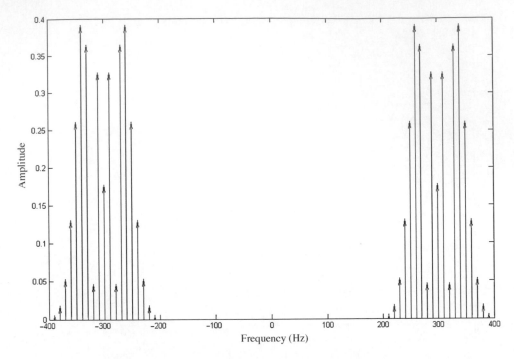

Figure 4-20   Amplitude spectrum of an FM signal with $\beta = 5$

bandwidth of an FM signal is theoretically infinite. However, the amplitude of the spectrum components of frequencies $f_c \pm nf_m$ is very small for large n. Hence we may define a finite effective bandwidth for the signal so that the bandwidth contains most, say 99%, of the total power.

Although the theoretical analysis indicates that the bandwidth of an FM system is quite large, there is an approximation rule suggested by Carson about it. Let W denote the bandwidth of the baseband signal. Let B_T denote the approximated bandwidth suggested by Carson. Recall that $\Delta f = k_f A_m$ as defined before. Then B_T is as follows:

$$B_T = 2\Delta f + W = 2k_m A_m + W. \tag{4-63}$$

This approximation rule is called *Carson's rule*.

The above discussion shows that the frequency modulation also lifts the baseband signal frequency to a higher carrier frequency f_c . The bandwidth of the modulated signal is infinite in theory and can be approximated by using Carson's rule as expressed in Equation (4-63).

Demodulation of an FM signal can be performed by taking the time derivation, called *slope detection*, followed by an envelope detector. A block diagram of such an FM demodulator is shown in Figure 4-21. Using Equation (4-43), the signal at the output of the differentiator can be expressed as

$$v(t) = -A_c[2\pi f_c + 2\pi k_f m(t)]\sin[2\pi f_c t + 2\pi k_f \int_{-\infty}^{t} m(\tau)d\tau]. \tag{4-64}$$

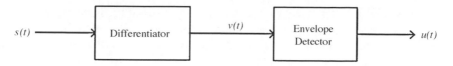

Figure 4-21   Block diagram of a slope detector for FM demodulation

The signal $v(t)$ has the following properties:

- As discussed in Section 4.1, the amplitude of

$$\sin[2\pi f_c t + 2\pi k_f \int_{-\infty}^{t} m(\tau)d\tau]$$

is $-A_c[2\pi f_c + 2\pi k_f m(t)]$, which is of the form $a + bm(t)$. This is again an envelope. Thus the envelope of $v(t)$ is as illustrated in Figure 4-18(d) for a sinusoidal message signal.
- Therefore, the output of the envelope detector contains a DC term proportional to the carrier frequency, namely $-A_c 2\pi f_c$, and a time-varying term proportional to the original message signal $m(t)$, namely $-A_c 2\pi f_c m(t)$. The DC term can be filtered out using a capacitor to obtain the desired demodulated signal.

The following discussion may be too advanced for computer science students. A demodulator using a differentiator is very easy to understand in mathematical terms, but in practice a differentiator may not be easy to implement. Actually, building a circuit that performs differentiation over the relevant band is sufficient.

Let us consider Figure 4-18 again. From 4-18(c) we can see that a high instantaneous frequency of the modulated signal corresponds to a large amplitude of the baseband signal. Therefore, it is desirable to have a circuit such that the gain is linearly proportional to the frequency of the input signal.

For example the band-pass *LCR* circuit shown in Figure 4-22(a) has frequency response shown in 4-22(b). The carrier frequency is placed at the middle of the linear region. The linear region of the filter acts like a differentiator. If the instantaneous frequency of the signal

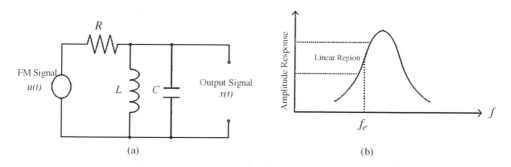

(a)   (b)

Figure 4-22   FM demodulator: (a) band-pass filter, and (b) the frequency response of the filter

is large, the gain of the filter is large and hence the output signal has large amplitude. On the other hand, if the instantaneous frequency of the signal is small, the gain of the filter is small and hence the output signal has small amplitude. Therefore, the message signal is transformed from frequency to amplitude. An AM detector is then used to extract the message signal from the envelope of the signal.

Let us consider the signal in Figure 4-18(c). If it goes through the band-pass filter described above, the output of the filter will be the signal shown in 4-18(d). As can be seen, the envelope of this output signal is indeed the baseband signal.

Perhaps the reader will be interested in knowing why the *RLC* circuit in Figure 4-22(a) is a band-pass filter. This is easy to see. Note that a high-frequency signal would effectively short-circuit the capacitor. Thus high-frequency signals will cause the output voltage to be

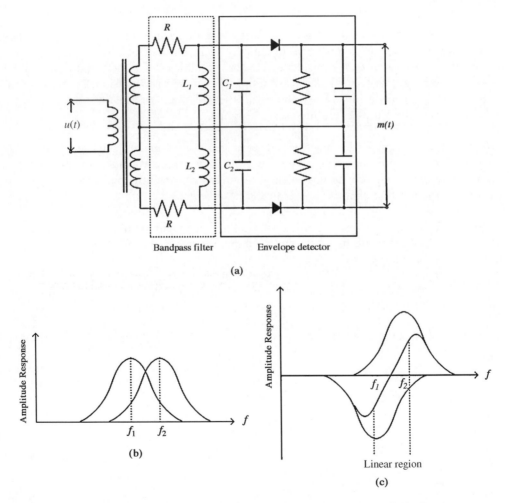

Figure 4-23   FM demodulator using a balanced discriminator: (a) the discriminator; (b) frequency response of the two high-pass filters; (c) the overall frequency response

zero, so they cannot pass through this circuit. Similarly, we can conclude that low-frequency signals will also effectively short-circuit the output terminals because of the inductance. Thus, both low-frequency and high-frequency signals are filtered out by the circuit. This is why this circuit is a band-pass filter.

A commonly used FM demodulator is the so-called *balanced discriminator*. The circuit is given in Figure 4-23(a). This FM demodulator uses two band-pass filters. One is placed at the upper branch with components R, L_1 and C_1, and the other is placed at the lower branch with components R, L_2 and C_2. The center frequencies of both filters are f_1 and f_2 which are slightly different. The frequency f_1 is controlled by L_1 and C_1, and f_2 is controlled by L_2 and C_2. Figure 4-23(b) shows the frequency responses of these filters. The outputs from both filters are then subtracted from each other, resulting the overall response given in 4-23(c). It is obvious that the linear region of the overall response is enlarged. The output signal is then demodulated by an envelope detector which produces the final message output.

4.5 Superheterodyne AM and FM Receivers

Radio broadcasting is a very familiar form of communication using wireless analog signal transmission. In this section we describe two types of broadcasting, AM and FM.

4.5.1 Superheterodyne AM Radio Receiver

The baseband message $m(t)$ for AM radio broadcasting is limited to a bandwidth of approximately 5 kHz; i.e. with $W = 5$ kHz. Amplitude-modulated radio broadcasting utilizes the frequency band 535–1605 kHz for transmission of voice and music. The carrier frequency allocations range from 540 to 1600 kHz with 10 kHz spacing. Since there are a huge number of receivers and relatively few radio transmitters, from an economic standpoint, the use of conventional AM for broadcasting results in very low cost receiver implementation.

A commonly used AM radio receiver is the so-called superheterodyne (or 'superhet') shown in Figure 4-26. The receiver consists of a radio-frequency (RF) amplifier, a mixer, a local oscillator, an intermediate-frequency (IF) amplifier, an envelope detector, an audio-frequency (AF) amplifier, and a loudspeaker. A mixer is a circuit whose inputs are two signals $f(t)$ and $g(t)$. Its output is $f(t)g(t)$; that is, a mixer produces the multiplication of two signals. Here, RF means high frequency, usually the carrier frequency. To tune for the desired radio frequency, two variable capacitors are used to tune the RF amplifier and the frequency of the local oscillator, respectively. In the superhet receiver, the signal received from the antenna after being amplified is converted to a common intermediate frequency $F_{IF} = 455$ kHz. The advantage of this conversion is that it allows the use of a single tuned IF amplifier for signals from any radio station in the frequency band. The IF amplifier is designed to have a bandwidth of 10 kHz; i.e. it is effectively band-pass filter with the passband from 450 to 460 kHz.

As we indicated before, the behavior of the mixer is similar to a multiplier which multiplies two signals. Its main function is to convert the carrier frequency of the received AM signal to IF. The frequency of the local oscillator is $F_{LO} = f_c + f_{IF}$, where f_c is the carrier frequency of the AM radio signal which we want to receive. The tuning range of the local oscillator is 955–2055 kHz. Let the received signal be $r(t) = A_c[1 + k_a m(t)]\cos(2\pi f_c t)$. By passing the received signal through the mixer, we have the following equation:

$$
\begin{aligned}
y(t) &= r(t)\cos(2\pi(f_c + f_{IF})t) \\
&= A_c[1 + k_a m(t)]\cos(2\pi f_c t)\cos(2\pi(f_c + f_{IF})t) \\
&= \frac{A_c[1 + k_a m(t)]}{2}\cos(2\pi f_{IF}t) + \frac{A_c[1 + k_a m(t)]}{2}\cos(2\pi(2f_c + f_{IF})t).
\end{aligned}
\tag{4-65}
$$

Therefore, we obtain two signal components, one centered at the intermediate frequency f_{IF} and one centered at the frequency $2f_c + f_{IF}$. Only the first component passes through the IF amplifier. The output of the IF amplifier is an AM signal with carrier frequency f_{IF}. This signal is then passed through an envelope detector as illustrated in Figure 4-3 which produces the desired audio message signal $m(t)$. Finally, the output of the envelope detector is amplified. The amplified audio signal then drives a loudspeaker.

Note that every amplifier in this circuit is a band-pass filter centered at a particular frequency. For a band-pass filter centered at f' with bandwidth W, only the signal with frequency components between $f' - W/2$ and $f' + W/2$ can pass the filter, and all signals with frequencies out of this band will be rejected. The RF amplifier is a band-pass filter centered at a particular RF frequency and the IF amplifier is also a band-pass filter centered at the IF frequency.

Why is the superheterodyne receiver is so important? If there is no IF stage, we will need an RF amplifier (band-pass filter) centered at frequency f_c with bandwidth 10 kHz. Since there can be roughly 100 such carrier frequencies, we have to be able to tune out around 100 such filters. Such a tunable and narrow band filter is not easy to implement in practice. With the superhet receiver, we will need an IF amplifier with a *fixed* center frequency of f_{IF} and a bandwidth of 10 kHz. For circuit implementation, a fixed narrow band-pass filter is relatively easy to implement compared with a tunable narrow one. In a superhet receiver, the RF filter can be less sophisticated, as will be explained.

Let us consider the receiver in Figure 4-24. Imagine that the RF filter is not very sophisticated, so that not only f_c goes through this amplifier, another carrier frequency, say

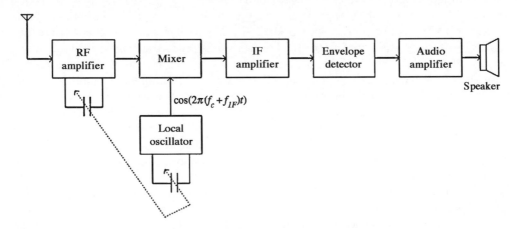

Figure 4-24 A superheterodyne AM radio receiver

$f_c + \Delta f$ where Δf is around 15 kHz, also goes through. That is, both carrier frequencies are amplified by the RF amplifier. Will the listener of the radio hear two broadcasts in this case? The answer is no. For the carrier frequency $f_c + \Delta f$, we will have the following signal after the mixer, namely

$$\frac{A_c(1 + k_a m'(t))}{2} \cos(2\pi(f_{IF} + \Delta f)t).$$

This signal cannot go through the IF amplifier, which is associated with an IF filter, as usually the IF filter has a bandwidth of 10 kHz.

Suppose $\Delta f = 2f_{IF}$. In this case, there will be another undesired signal, namely

$$r'(t) = \frac{A_c(1 + k_a m'(t))}{2} \cos(2\pi(f_c + 2f_{IF})t),$$

together with the desired signal passing through the RF amplifier. Consider

$$y'(t) = r'(t)\cos(2\pi(f_c + f_{IF})t)$$
$$= \frac{A_c(1 + k_a m'(t))}{2} \cos(2\pi f_{IF}t) + \frac{A_c(1 + k_a m'(t))}{2} \cos(2\pi(?f_c + 3f_{IF})t)$$

It can now be seen that the undesired signal will also go through the IF amplifier, which is not what we want. Thus, the RF filter will have to be able to filter out this kind of signal. That is, if the RF filter is tuned to the center frequency f_c, the maximum bandwidth of the RF filter can be $4f_{IF}$ and the frequency $f_c + 2f_{IF}$ can still be rejected. But this is quite wide as f_{IF} is 455 kHz. So the RF filter needs to have a bandwidth roughly equal to 1820 kHz, which is much easier to design than a filter with 10 kHz.

4.5.2 Superheterodyne FM Radio Receiver

The frequency band utilized for FM radio broadcasting is 88–108 MHz. The carrier frequencies are separated by 200 kHz and the peak-frequency deviation is fixed at 75 kHz.

Superheterodyne receivers are commonly used in FM radio broadcasting. The block diagram of such a receiver is shown in Figure 4-25. To tune for the desired radio frequency, two variable capacitors are used to tune the RF amplifier and the frequency of the local oscillator. This allows the mixer to bring any one of all possible FM radio signals to a common IF frequency of $f_{IF} = 10.7$ MHz with bandwidth 200 kHz. Let the received signal be

$$r(t) = A_c \cos\left[2\pi f_c + 2\pi k_f \int_{-\infty}^{t} m(\tau)d\tau\right]. \tag{4-66}$$

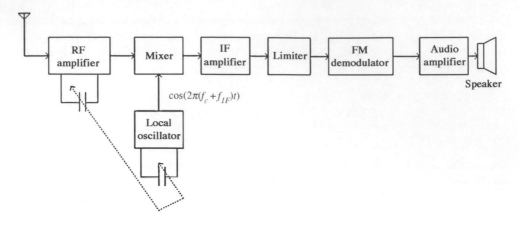

Figure 4-25 A superheterodyne FM radio receiver

By passing the received signal through the mixer, we have the following equation:

$$
\begin{aligned}
y(t) &= r(t)\cos[2\pi(f_c + f_{IF})t] \\
&= A_c\cos[2\pi f_c + 2\pi k_f \int_{-\infty}^{t} m(\tau)d\tau]\cos[2\pi(f_c + f_{IF})t] \\
&= \frac{A_c}{2}\cos[2\pi f_{IF}t - 2\pi k_f \int_{-\infty}^{t} m(\tau)d\tau] \\
&\quad + \frac{A_c}{2}\cos[2\pi(2f_c + f_{IF})t + 2\pi k_f \int_{-\infty}^{t} m(\tau)d\tau].
\end{aligned}
\tag{4-67}
$$

Therefore, we obtain two signal components, one centered at the intermediate frequency f_{IF} and one centered at the frequency $2f_c + f_{IF}$. Again, only the first component is passed by the IF amplifier. A band-pass filter centered at $f_{IF} = 10.7$ MHz with a bandwidth 200 kHz is employed to remove unwanted signals. The output of the IF amplifier is an FM signal with carrier frequency f_{IF}. The amplitude limiter removes any amplitude variation in the received signal at the output of the IF amplifier. A balance frequency discriminator is used for frequency demodulation. The resulting message signal is passed to the audio amplifier. The amplified signal is then used to drive a loudspeaker.

4.5.3 Walkie-talkie and Amateur Radio

Both walkie-talkie and amateur radio systems use the AM and FM technologies.

A walkie-talkie is a hand-held portable, bidirectional radio transceiver. In the early days, walkie-talkie operated in the 27 Mhz frequency band and used amplitude modulation only. Later it transfered to the 49 MHz band, sometimes with frequency modulation.

Amateur radio (also called ham radio) is a hobby public service. Usually, only voice is transmitted by ham radios. The modulation techniques employed are frequency modulation or single-sideband. The frequency range is from 1.8 MHz to 440 MHz.

4.6 Analog Modulation with Frequency Division Multiplexing

An important signal processing approach that employs analog modulation techniques is frequency division multiplexing. This technique allows a number of independent signals to be combined into a composite signal suitable for transmission over a common channel. The frequency of each signal must be elevated to a distinct carrier frequency so that it does not interference with any other signal, and then the signals can be separated at the receiving end. This can be achieved by analog modulation because analog modulation always transforms the frequencies of the baseband signal into the carrier frequency band.

Different signals with the same baseband frequencies now have distinct carrier frequencies. We may say that signals are now effectively separated by frequencies. This technique is called frequency-division multiplexing (FDM).

A block diagram of an FDM system is shown in Figure 4-26. The N message inputs are assumed to be of the low-pass type. Each message input is passed through a low-pass filter to remove high-frequency components that do not contribute significantly to signal representation but are capable of disturbing other modulated signals that share the common channel. If the message signal is originally band-limited, the low-pass filters may be omitted. The filtered signals are then applied to the analog modulators that shift the frequency ranges of the signals to occupy mutually exclusive carrier frequency intervals. Each carrier signal is obtained from a carrier supply. The analog modulation method in this system may be one of the techniques described earlier. The output signals from the modulators are then filtered by a bank of band-pass filters so that the frequency bands of all the modulated signals are mutually exclusive strictly. The resulting band-pass modulated signals are then combined and transmitted through the common channel. At

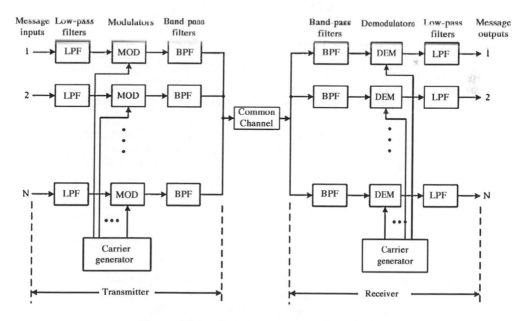

Figure 4-26 Block diagram of an FDM system

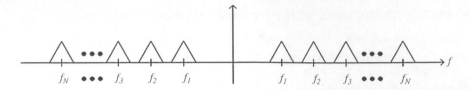

Figure 4-27   Spectrum of a combined FDM signal

the receiving end, the received signal is the combination (addition) of all the transmitted modulation signals.

Figure 4-27 illustrates a possible spectrum of the signal that corresponds to the combination of all modulated signals. At the receiver, a bank of band-pass filters are then used to separate the message signals so that each output of the band-pass filter corresponds to the original modulated waveform. Finally, the message signals are recovered by individual demodulators followed by low-pass filters. To provide for two-way transmission, as in analog mobile phone systems, we have to completely duplicate the system shown in Figure 4-26 with the signal direction and components connected in reverse order.

Let us use a typical cable TV system as an example. The cable may have a bandwidth of 500 MHz. The lowest frequency may be around 50 MHz and the highest may be 550 MHz. The bandwidth of each color TV channel is roughly 8 MHz. Thus we can see that this cable TV system can provide roughly 60 TV channels. The analog modulation technique is very useful in this case as each TV station must use a unique carrier frequency allocated by the system.

In the above, we have assumed that the TV stations broadcast analog signals. If digital TV broadcasting is available, digital modulation is used. This will be introduced in the next chapter. The reader will find that analog modulation is still important and useful there.

4.7 Concluding Remarks

In this chapter we have introduced analog modulation techniques. Essentially, analog modulation performs two functions.

First, since voice and music signals are all low-frequency signals, they cannot be transmitted directly in a wireless fashion. This is due to the fact that the length of an antenna is $\lambda/2$ and $\lambda = v/f$, where λ is the wavelength, v is the velocity of light and f is the frequency. The wavelength of a voice signal is therefore very large and no antenna can be built with such a large length. Analog modulation lifts the frequencies in a baseband signal to the carrier frequency band. Since the carrier frequency is large, the wavelength becomes small enough for an antenna to be built. It is customary to call carrier frequencies 'radio frequencies'. Therefore, we often hear the term radio-frequency technology which always involves analog modulation.

Second, in a multiplexing environment, many transmitters will transmit their individual signals through the same channel. Analog modulation makes this possible by giving each transmitter a distinct carrier frequency. The receiver uses band-pass filters to extract each individual signal, now identified by its carrier frequency.

Further Reading

- Analog communication systems are treated in numerous books on basic communications theory, including [S79], [C86], [S90], [C93], [G93], [H00] and [ZT02].
- Carson's rule is named in honor of its originator. Carson and Fry wrote an important paper on frequency modulation theory; see [CF37].
- For a more detailed description of the superheterodyne receiver, see [H59].

Exercises

4.1 Use four different methods to explain why amplitude modulation would lift the frequencies of the base band to higher ones.

4.2 Explain the major concept of demodulation in double sideband.

4.3 Explain why single sideband works.

4.4 In practice, discrete Fourier transform, instead of Fourier transform, is used. Describe how the discrete Fourier transform can be used to implement the single sideband.

4.5 Let $x(t) = \cos\{(2\pi t/4) + (\pi/6)\}$ and $n = 4$. Find its discrete Fourier transform. Show that by applying the techniques covered in question 4 above, you may obtain $x'(t) = \sin\{(2\pi t/4) + (\pi/6)\}$ by modifying the coefficients of the discrete Fourier transform of $x(t)$ and applying the inverse discrete Fourier transform to them.

4.6 If analog modulation is not used, can we transmit human voice wirelessly?

4.7 If analog modulation is not used, can we distinguish between different radio stations?

5

Digital Modulation Techniques

In the previous chapter, we introduced methods that modulate a continuous function (message signal) onto a carrier by varying the carrier's amplitude or phase. This chapter introduces digital modulation techniques. The main difference between analog and digital modulation is that the transmitted message for a digital modulation system represents a small set of abstract symbols (e.g. 0s and 1s for a binary transmission system), while in an analog modulation system the message signal is a continuous waveform. To transmit a digital message, digital modulation allocates a piece of time called a *signal interval* and generates a continuous function that represents the symbol.

For digital modulation, the message signal is often transformed into a baseband signal. In a wireless communications system, a second part of the modulator converts the baseband signal to a radio-frequency (RF) signal, modulating the phase, frequency or amplitude of the carrier. In a wired system, the baseband signal may be sent directly without carrier modulation. However, sometimes, multiple message signals are required to transmit simultaneously through the same wire. Under these circumstances, some modulation techniques may be employed so that different messages can be modulated into different frequency bands. This technique is called *frequency-division multiple access* (FDMA). Besides this, there are more multiple access techniques that may be employed for digital systems, such as *time-division multiple access* (TDMA) and *code-division multiple access* (CDMA). Those techniques will be introduced in the next chapter.

The receiver end of a digital modulation system consists of a circuit to convert the modulated signal to a baseband signal and circuits to decide which symbol is transmitted during each signal interval by the transmitter. A digital demodulator differs from an analog one in that it outputs a specified symbol, whereas an analog demodulator produces an output that approximately equals the message signal. These facts are important in deciding whether to use a digital modulation scheme.

One may easily confuse digital conversion with digital modulation. The former changes the form of information from analog to digital. Digital *modulation* is required when we want to transmit analog signals via a digital communication system but is unnecessary if the information is already in digital form.

The major purpose of this chapter is to focus on methods used in various digital modulation and demodulation schemes, not on analysis of the performance of those systems (which in general requires more mathematical background, such as probability theory and

Communications Engineering.   R.C.T. Lee, Mao-Ching Chiu and Jung-Shan Lin
© 2007 John Wiley & Sons (Asia) Pte Ltd

stochastic processes). The performance of those systems will be explained intuitively whenever possible.

5.1 Baseband Pulse Transmission

In this section we will introduce a transmission technique which does not require carrier modulation, called *baseband pulse transmission*. The digital information is transformed into a pulse train. We will restrict the discussion to the binary transmission case. Each pulse has a duration and represents specific digital information. In this form of signaling, symbols 1 and 0 are represented by positive and negative rectangular pulses of equal amplitude and equal duration. For example, if the pulse considered is a rectangular pulse, the binary information stream of 1001101 is transformed into a baseband signal as shown in Figure 5-1(a), where T_b is assumed to be one second.

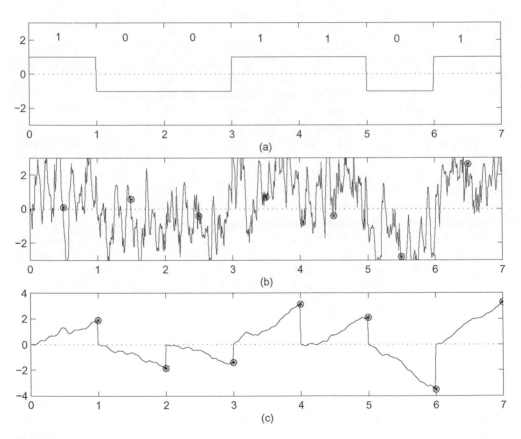

Figure 5-1   Signals for baseband pulse transmission: (a) baseband pulse signal; (b) the received signal corrupted by noise, and the sampled points; (c) the output of the correlator and the corresponding sampling points

In mathematical notation, letting m be the transmitted symbol (i.e. $m \in \{0, 1\}$), we have the transmitted signal denoted as

$$s(t) = \begin{cases} +A & \text{if} \quad m = 1 \\ -A & \text{if} \quad m = 0 \end{cases} \qquad (5\text{-}1)$$

for $0 \leq t \leq T_b$. In this case, $1/T_b$ is called the *bit rate*. Note that, for mathematical representation, it is sufficient to consider the transmission over $0 \leq t \leq T_b$, since similar cases happen if a different signaling interval is considered.

It is assumed that the receiver has acquired knowledge of the starting and ending times of each transmitted pulse; in other words, the receiver has prior knowledge of the pulse shape, but not its polarity. Given the received signal, the receiver is required to make a decision in each signaling interval as to whether the transmitted symbol is a 1 or a 0.

At the receiving end, the simplest way of recovering the original digital stream is to sample the received signal at the sampling rate $1/T_b$. Then a decision device is used to 'guess' the transmitted symbol based on the sampled value at each instance. The sampling instance is in general chosen to be in the middle of the signaling interval. If the sampled value is positive, it decides that a 1 was transmitted. If the sampled value is negative, it decides that a 0 was transmitted. This scheme was used as standard in short-distance wired transmission (e.g. the RS232 standard). However, for long-distance transmissions, noise will be added to the transmitted signal. In addition, the desired transmitted signal will be seriously attenuated over long distances. As a result, the received signal will not be 'clean' compared with the original transmitted signal. For example, Figure 5-1(b) gives a noisy version of the received signal and the corresponding sampled points which are indicated by circles. In this case, if a sampled value is directly used to decide which symbol was transmitted by the transmitter, it is very possible that the sampled value goes to the opposite polarity at the sampling instance. The decision device will make a wrong decision based on this sampled value. As shown in Figure 5-1(b), the decisions made based on the sampled values produce the output 1101001, which contains two errors. Therefore, this scheme is not suitable for long-distance transmission.

In the following, we will explain the meaning of Figure 5-1(c). If only one sampled value is used by the decision device, the whole waveform over $(0, T_b)$ is observed at only one instance. This scheme is obviously not optimal, since we do not take advantage of everything known about the signal. Since the starting and ending times of the pulse are known, a better procedure is to compare the area of the received signal-plus-noise waveform by integrating the received signal over the T_b-second signaling interval.

A receiver for baseband pulse transmission is given in Figure 5-2, where an integrator is employed. The integrator will integrate the waveform over $(0, T_b)$ with the output being sampled at time T_b. Of course, a noise component is present at the output of the integrator; but, since the additive noise is always assumed to be of zero mean, it takes on positive and

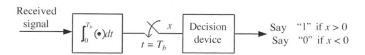

Figure 5-2 Receiver for baseband pulse transmission

negative values with equal probability. Thus the output noise component has zero mean. Intuitively, the integrator can be thought of as an energy collector that collects the energy of the received signal over $(0, T_b)$, producing an energy value with sign at the sampling instance. In addition, the noise will be 'averaged' over $(0, T_b)$ by the integrator, and hence the noise component will be suppressed by the integrator. Formally, the integrate-and-dump device is called a **correlator**.

The decision device then, based on the sampled values, makes a decision on the transmitted symbol. If the output of the correlator at the sampling instance is positive, it decides that a 1 was transmitted. If the output is negative, it decides that a 0 was transmitted. Figure 5-1(c) gives the output waveform of the integrator, with the sampling points indicated by circles. The decisions based on these sampled values are 1001101, which contains no error.

It is very important to determine the required bandwidth for the transmitted signal. For example, for broadband transmission, if the signal has a bandwidth which is greater than the bandwidth of the channel, the transmitted signal will not be able to transmit through the channel without distortion.

For digital communication, the digital signals are transmitted consecutively one-by-one. Therefore, we may express the transmitted signal as

$$s(t) = \sum_{k=-\infty}^{\infty} b_k h(t - kT_b), \tag{5-2}$$

where $b_k \in \{+1, -1\}$ for binary transmission and $h(t)$ is a pulse shaping function. In our case, we may let

$$h(t) = \begin{cases} 1 & 0 \leq t < T_b \\ 0 & \text{otherwise.} \end{cases} \tag{5-3}$$

5.1.1 Power Spectral Density

In communication systems, we have to know how the transmitted power is distributed over the frequencies. For example, we may need to know whether most of the transmitted power passes through the channel so that the received signal can have better signal quality. Now the question is: What is the power distribution of the signal $s(t)$ in Equation (5-2)? To answer this, we must first introduce the concept of *power spectral density*. A formal and rigorous explanation requires knowledge of random processes. However, since the meaning of the power spectral density of $s(t)$ is similar to the Fourier transform of $s(t)$, we may explain it intuitively.

Let $s_T(t)$ be a truncated signal of $s(t)$ given by

$$s_T(t) = \begin{cases} s(t) & -T/2 \leq t \leq T/2 \\ 0 & \text{otherwise.} \end{cases} \tag{5-4}$$

The power spectral density for $s(t)$ is defined as

$$P_s(f) = \lim_{T \to \infty} \frac{1}{T} |S_T(f)|^2, \tag{5-5}$$

where $S_T(f)$ is the Fourier transform of $s_T(t)$.

The power of the signal $s(t)$ is defined as

$$P_s = \lim_{T \to \infty} \frac{1}{T} \int_{-T/2}^{T/2} |s(t)|^2 dt. \tag{5-6}$$

A very useful theorem, called **Parseval's theorem**, is as follows:

$$\int_{-T/2}^{T/2} |s(t)|^2 dt = \int_{-\infty}^{\infty} |S_T(f)|^2 df. \tag{5-7}$$

Therefore we have

$$P_s = \lim_{T \to \infty} \frac{1}{T} \int_{-T/2}^{T/2} |s(t)|^2 dt$$

$$= \lim_{T \to \infty} \frac{1}{T} \int_{-\infty}^{\infty} |S_T(f)|^2 df$$

$$= \int_{-\infty}^{\infty} \left(\lim_{T \to \infty} \frac{|S_T(t)|^2}{T} \right) df. \tag{5-8}$$

Substituting Equation (5-5) into (5-8), we have

$$P_s = \int_{-\infty}^{\infty} P_s(f) df. \tag{5-9}$$

Therefore, the power of $s(t)$ is exactly equal to the integration of the power spectral density over frequency. This is why $P_s(f)$ is called the power spectral density of $s(t)$. Now our question is how to find $P_s(f)$.

From Equation (5-2) we can see that the signal $s(t)$ is a combination of the same waveform $h(t)$ of different delays and amplitudes. Let $H(f)$ be the Fourier transform of $h(t)$. The power spectral density $P_s(f)$ of $s(t)$ can be shown to be

$$P_s(f) = \frac{1}{T_b} |H(f)|^2. \tag{5-10}$$

Since, in this case, $h(t)$ is a rectangular pulse, according to Equation (3-102), the Fourier transform of $h(t)$ is

$$H(f) = e^{-j\pi f T_b} \frac{\sin(\pi f T_b)}{\pi f} = e^{-j\pi f T_b} T_b \text{sinc}(f T_b), \tag{5-11}$$

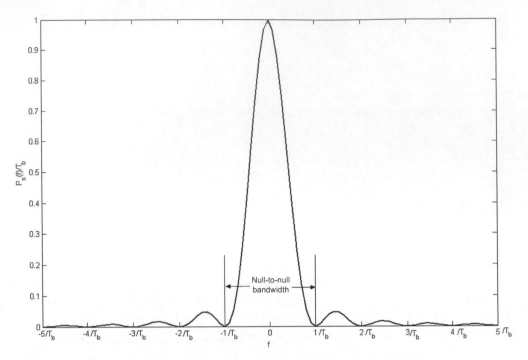

Figure 5-3 Power spectral density of the transmitted baseband signal $s(t)$

where

$$\operatorname{sinc}(x) = \frac{\sin(\pi x)}{\pi x}. \tag{5-12}$$

Therefore we have $|H(f)| = T_b\operatorname{sinc}(fT_b)$ and the power spectral density $P_s(f)$ of $s(t)$ is

$$P_s(f) = T_b\operatorname{sinc}^2(fT_b). \tag{5-13}$$

Figure 5-3 plots the power spectral density of $s(t)$ with $T_b = 1$. It can be seen that most of the power of $s(t)$ concentrates over $[-1/T_b, 1/T_b]$. Since the power spectral density is zero at $-1/T_b$ and $1/T_b$, we define the bandwidth of the baseband signal, also called the *null-to-null bandwidth*, as $2/T_b$.

Several points about the above discussion are in order.

- In the above, we used $|X(f)|^2$ instead of $X(f)$. This appears to be different from what we did in Chapter 4. Actually, we used this concept implicitly before. In Chapter 4 we pointed out that $X(f)$ is a complex number in the general case. Therefore, when we presented the Fourier transform spectrum, we actually were using $|X(f)|^2$. That is, if $X(f) = a + jb$, we were showing $a^2 + b^2$.

We should be analyzing the spectrum of a sequence of pulses. Instead, we only analyze the spectrum of a signal pulse. This is due to the fact that each digital signal is represented by a random sequence of 1s or 0s. Thus, it suffices to analyze one pulse, instead of a sequence.

5.1.2 Data Rate and Signal Bandwidth

It is very important to identify the differences between *data rate* and *signal bandwidth* in a digital baseband transmission system.

- The data rate is defined as the number of bits transmitted per second. For example, a data rate of 1 Mbits/s means that the system is capable of transmitting 1 megabits every second.
- The signal bandwidth is defined as the bandwidth in which most of the signal power is concentrated. For example, we may define the bandwidth of a signal as the width of the spectrum in which 99% of the power of the signal is concentrated. Or, we may simply use the null-to-null bandwidth. For example, from Figure 5-3, the null-to-null bandwidth of the transmitted signal is $2/T_b$.

It can be seen that the bandwidth is proportional to the bit rate. If the bit rate is high, T_b is small and the bandwidth becomes very large. This means that the communication system handling high data rate transmission must be able to cope with a wide range of frequencies.

Let us consider the situation where multiple users are using the channel. In such a case, the baseband pulse transmission method does not work because there is no way to distinguish these users. Further, if the signals are transmitted in a wireless fashion, this method does not work because electromagnetic waves are sinusoidal signals. These problems will all disappear when we introduce carrier waves into the system in the following sections.

5.2 Amplitude-shift Keying (ASK)

From now on, we will add carrier waves into the modulation scheme. The simplest modulation technique is the amplitude shift keying (ASK), where the digital information is modulated over the carrier's amplitude, similar to the amplitude modulation for analog modulation.

In general, we require two waveforms $s_1(t)$ and $s_2(t)$ for binary transmission. If the transmitter wants to transmit a 1, $s_1(t)$ is employed over the signaling interval $(0,T_b)$. On the other hand, if the transmitted symbol is a 0, $s_2(t)$ is used over $(0,T_b)$. For ASK, the transmitted waveforms can be expressed as

$$s_1(t) = \sqrt{\frac{4E_b}{T_b}} \cos(2\pi f_c t)$$
$$s_2(t) = 0$$

(5-14)

for $0 \leq t \leq T_b$, where E_b is the averaged transmitted signal energy per bit and f_c is the carrier frequency which is equal to n_c/T_b for some fixed integer n_c. The averaged transmitted energy is expressed as follows:

$$0.5 \int_0^{T_b} s_1^2(t)dt + 0.5 \int_0^{T_b} s_2^2(t)dt.$$

(5-15)

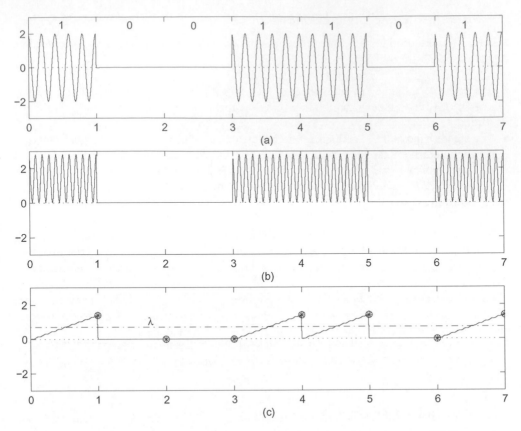

Figure 5-4 Signals for ASK: (a) transmitted signal; (b) the signal $y(t) = s(t) \times \sqrt{2/T_b}\,\cos(2\pi f_c t)$; (c) the output of the integrator and the corresponding sampling points

The transmitted signal $s(t)$ can be expressed as

$$s(t) = \begin{cases} s_1(t) & \text{for symbol} \quad 1 \\ s_2(t) & \text{for symbol} \quad 0 \end{cases} \qquad (5\text{-}16)$$

for $0 \leq t \leq T_b$. We note that the transmitter for such a system simply consists of an oscillator that is gated on and off, and accordingly ASK is often referred to as *on–off keying*. Figure 5-4(a) gives a transmitted waveform resulting from the digital transmission of 1001101, where $E_b = 1$, $f_c = 5\,\text{Hz}$, and $T_b = 1\,\text{s}$.

A receiver for ASK is given in Figure 5-5. In the following, we will explain how the demodulation works. The received signal is first multiplied by a unit-energy signal $\sqrt{2/T_b}\,\cos(2\pi f_c t)$. Assuming that the received signal is noise-free, we have the following expression after multiplication:

$$y(t) = s(t) \times \sqrt{\frac{2}{T_b}}\,\cos(2\pi f_c t). \qquad (5\text{-}17)$$

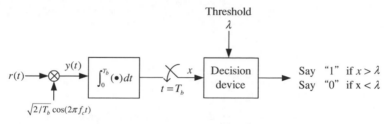

Figure 5-5 A receiver for amplitude-shift keying

Figure 5-4(b) shows the corresponding signal $y(t)$. If

$$s(t) = \sqrt{\frac{4E_b}{T_b}} \cos(2\pi f_c t) \tag{5-18}$$

we have

$$y(t) = \frac{\sqrt{8E_b}}{T_b} \cos^2(2\pi f_c t). \tag{5-19}$$

Thus $y(t) \geq 0$ for the corresponding period. An integrator is then employed to integrate the signal $y(t)$ over $(0, T_b)$ with the output being sampled at time T_b. Defining the sampled signal as x, we have

$$
\begin{aligned}
x &= \int_0^{T_b} \sqrt{\frac{4E_b}{T_b}} \cos(2\pi f_c t)\sqrt{\frac{2}{T_b}} \cos(2\pi f_c t)dt \\
&= \frac{\sqrt{8E_b}}{T_b} \int_0^{T_b} \frac{1 + \cos(4\pi f_c t)}{2} dt.
\end{aligned}
\tag{5-20}
$$

Replacing $f_c = n_c/T_b$ in the above equation, we have

$$
\begin{aligned}
x &= \frac{\sqrt{8E_b}}{T_b} \left(\int_0^{T_b} \frac{1}{2}dt + \int_0^{T_b} \frac{1}{2}\cos\left(4\pi \frac{n_c}{T_b}t\right)dt \right) \\
&= \frac{\sqrt{8E_b}}{T_b} \left(\frac{1}{2}t \Big|_0^{T_b} + \frac{1}{2}\frac{1}{4\pi(n_c/T_b)}\sin\left(4\pi\frac{n_c}{T_b}t\right)\Big|_0^{T_b} \right) \\
&= \sqrt{2E_b}.
\end{aligned}
\tag{5-21}
$$

The last equality follows since the second term in the brackets is zero. If $s_2(t)$ was transmitted, the sampled signal is $x = 0$ since $s_2(t) = 0$. In summary, we have

$$
x = \begin{cases} \sqrt{2E_b} & \text{if 1 is transmitted} \\ 0 & \text{if 0 is transmitted.} \end{cases}
\tag{5-22}
$$

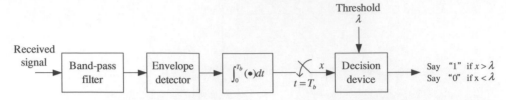

Figure 5-6   Noncoherent detector for amplitude-shift keying

Figure 5-4(c) gives the resulting signal from the integrator and the corresponding sampled points. The decision device based on the sampled values makes a decision of 1001101 which is exactly the same as the transmitted sequence.

Although there are only two possible outcomes for the noiseless case, the actual received signal may be corrupted by noise. For certain systems, we have infinitely many outcomes for the output of the correlator. Therefore, a threshold value λ is required in the decision device to distinguish which symbol was transmitted. This value is set to the center of 0 and $\sqrt{2E_b}$, which is $\lambda = \sqrt{E_b/2}$, so that minimum bit error probability can be achieved. If $x > \lambda$, the decision device makes a decision of 1. If $x < \lambda$, the decision device makes a decision of 0.

The detection method described requires a carrier with the same frequency and phase as those of the carrier in the transmitter. These types of receiver are called ***coherent detectors***. It is also possible to detect an ASK signal without exact knowledge of the carrier frequency and phase. Those detectors are called ***noncoherent detectors***.

For binary ASK, a noncoherent detector is illustrated in Figure 5-6 which consists of a band-pass filter which will pass the input signal and an envelope detector followed by a correlator and a decision device. The band-pass filter is used to remove the out-of-band noise. The envelope detector is then used to recover the envelope of the carrier wave. The resulting signal is a baseband pulse signal which can then be detected by the baseband pulse detector, a correlator followed by a decision device as described in the previous section. In general, it can be proved that the transmitted energy for the noncoherent detector must be doubled to achieve the same bit error probability as that of the coherent detector. The proof is omitted in this book.

That we may have a noncoherent detector for ASK is due to the fact that the ASK system is a one or nothing system. Note that when a 1 is sent there is signal and when 0 is sent there is no signal. Therefore, we only have to detect whether there is a signal or not. That is why we may have the noncoherent detector for ASK.

In the following, a geometric interpretation is used to represent a signal set (e.g. $\{s_1(t), s_2(t)\}$ for the binary case) as points in the signal space. The collection of all possible signal points is called the *signal constellation*. For example, for the binary ASK, let

$$\phi_1(t) = \sqrt{\frac{2}{T_b}} \cos(2\pi f_c t), \tag{5-23}$$

which is a unit energy signal over $(0, T_b)$. Then, $s_i(t)$ for $i = 1, 2$ can be represented as

$$s_i(t) = s_{i1} \phi_1(t). \tag{5-24}$$

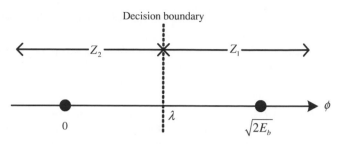

Figure 5-7 Geometric representation of binary ASK signals

It is easy to show that $s_{11} = \sqrt{2E_b}$ and $s_{21} = 0$. The geometric interpretation of these two signal points can be denoted as points in a real line as illustrated in Figure 5-7. Note that, according to Equation (5-22), the received signals after demodulation are either $\sqrt{2E_b}$ or 0, just the same as the transmitted signals. The decision boundary is therefore determined by the threshold value λ. If the received signal x lies on the right side of the decision boundary (i.e. the region Z_1), then a decision of a 1 is made. If the received signal x lies on the left side of the decision boundary (i.e. the region Z_2), then a decision of a 0 is made.

One advantage in using the signal space representation is that it is much easier to identify the 'distance' between signal points. The distance is closely related to the symbol error rate of a given constellation. In this ASK case, the distance between two signal points is $\sqrt{2E_b}$. Intuitively, to reduce the probability of detection error due to noise, we may increase the transmitted energy of the signals; i.e. increase E_b. As a result, the distance between two signal points will be increased, which makes the received signal point less likely to be located in the wrong region. Note that increasing E_b is equivalent to increasing the magnitude of the signal $s_1(t)$, so more power is needed to transmit the signals.

Sometimes, multi-dimensional geometric interpretation is required to represent the whole signal set over the signaling interval $(0, T)$. For the N-dimensional case, a set of orthonormal functions $\{\phi_1(t), \phi_2(t), \ldots, \phi_N(t)\}$ is required to present all the possible transmitted signals $\{s_1(t), s_2(t), \ldots, s_M(t)\}$. Note that $M = 2$ for the binary case. The set of functions $\{\phi_1(t), \phi_2(t), \ldots, \phi_N(t)\}$ forms an orthonormal basis so that every transmitted signal can be presented as a linear combination of the basis functions. That is:

$$s_i(t) = \sum_{j=1}^{N} s_{ij}\phi_j(t) \qquad \text{for } i = 1, \ldots, M \text{ and } 0 \leq t \leq T \qquad (5\text{-}25)$$

and

$$\int_0^T \phi_i(t)\phi_j(t)dt = \begin{cases} 1 & i = j \\ 0 & i \neq j. \end{cases} \qquad (5\text{-}26)$$

We may plot the signal vector $s_i = [s_{i1}, s_{i2}, \cdots, s_{iN}]$ as a point in the N-dimensional signal space to represent the time-domain signal $s_i(t)$. These concepts will become clear as we introduce multi-dimensional modulations: *frequency-shift keying* (FSK) and *quadriphase-shift keying* (QPSK).

Note that the digital information is now transmitted as analog signals. In wireless communication systems, only analog signals can be transmitted. When the signal is transmitted, its carrier frequency is f_c. As for the bandwidth, this will be discussed in the next section.

Since carrier frequencies are now used, the multi-user problem can be solved. Every user is given a distinct carrier frequency f_c. The receiver will use a band-pass filter to obtain signals with this particular frequency. If there are K users, K such bandpass filters are needed.

If the carrier frequency f_c is not high enough, one may use the analog modulation technique to up-convert it to an even higher radio frequency (RF).

5.3 Binary Phase-shift Keying (BPSK)

In a binary phase-shift keying system, we use a pair of signals $s_1(t)$ and $s_2(t)$ to represent binary symbols 1 and 0, respectively, as follows:

$$s_1(t) = \sqrt{\frac{2E_b}{T_b}} \cos(2\pi f_c t) \tag{5-27}$$

$$s_2(t) = \sqrt{\frac{2E_b}{T_b}} \cos(2\pi f_c t + \pi) = -\sqrt{\frac{2E_b}{T_b}} \cos(2\pi f_c t), \tag{5-28}$$

for $0 < t < T_b$, where E_b is the transmitted energy per bit, and f_c is the carrier frequency which is chosen to be equal to n_c/T_b for some fixed integer n_c. Again, as in the ASK case, the averaged transmitted signal energy which is expressed as

$$0.5 \int_0^{T_b} s_1^2(t)dt + 0.5 \int_0^{T_b} s_2^2(t)dt$$

is equal to E_b. A pair of sinusoidal waves that differ in a relative phase shift of 180 degrees are referred to as antipodal signals. Figure 5-8(a) illustrates a transmitted signal when the binary stream is 1001101, where $E_b = 1, f_c = 5$ Hz and $T_b = 1$ s.

An optimal detector for the BPSK system is shown in Figure 5-9. The received signal is first multiplied by the unit energy **truncated** signal $\phi_1(t) = \sqrt{2/T_b} \cos(2\pi f_c t)$ if $0 \leq t \leq T_b$, or by $\phi_1(t) = 0$ otherwise. Assuming that the received signal is noise-free, we have the following expression after multiplication

$$y(t) = s(t) \times \sqrt{\frac{2}{T_b}} \cos(2\pi f_c t). \tag{5-29}$$

Therefore we have

$$y(t) = \begin{cases} \dfrac{2\sqrt{E_b}}{T_b} \cos^2(2\pi f_c t) & \text{if 1 was transmitted} \\[4mm] -\dfrac{2\sqrt{E_b}}{T_b} \cos^2(2\pi f_c t) & \text{if 0 was transmitted.} \end{cases} \tag{5-30}$$

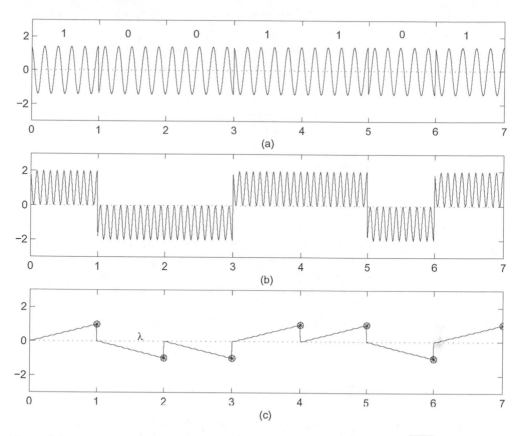

Figure 5-8 Signals for BPSK. (a) transmitted signal, (b) the signal $y(t) = s(t)\sqrt{2/T_b}\cos(2\pi f_c t)$; (c) the output of the correlator and the corresponding sampling points

An example of the signal $y(t)$ is plotted in Figure 5-8(b). The two possible outcomes from the output of the correlator for the noiseless case can be shown to be

$$x = \begin{cases} +\sqrt{E_b} & \text{if 1 was transmitted} \\ -\sqrt{E_b} & \text{if 0 was transmitted.} \end{cases} \tag{5-31}$$

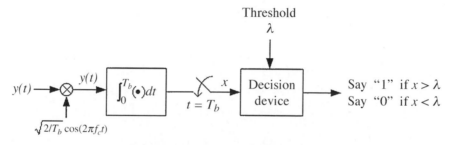

Figure 5-9 Coherent detector for BPSK

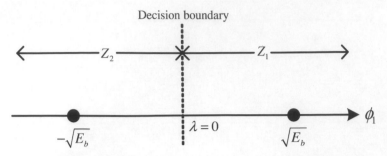

Figure 5-10 Signal-space diagram for coherent BPSK

The threshold value λ in this case must be set to 0 which corresponds to the center between $-\sqrt{E_b}$ and $\sqrt{E_b}$. The output signal of the integrator and the corresponding sampling points are plotted in Figure 5-8(c). If the sampled value is greater then zero, a decision of 1 is made. If the sampled value is less than zero, a decision of 0 is made. The decision device based on the sampled values makes a decision of 1001101, which is exactly the same as the transmitted sequence.

For geometric representation, the transmitted signals $s_1(t)$ and $s_2(t)$ can be expressed in terms of $\phi_1(t) = \sqrt{2/T_b}\cos(2\pi f_c t)$ as follows:

$$\begin{aligned} s_1(t) &= \sqrt{E_b}\phi_1(t), \qquad 0 \le t \le T_b \\ s_2(t) &= -\sqrt{E_b}\phi_1(t), \qquad 0 \le t \le T_b. \end{aligned} \tag{5-32}$$

A coherent BPSK system is therefore characterized by a signal space that is one-dimensional, with a signal constellation consisting of two message points as shown in Figure 5-10.

Again, as seen from Equation (5-31), the received signals after demodulation are the same as the transmitted signals, and the decision boundary shown in Figure 5-10 is determined by the threshold value λ, which in this case is equal to 0. If the received signal x lies on the right side of the decision boundary (region Z_1), then a decision of 1 is made. If the received signal lies on the left side of the decision boundary (region Z_2), then a decision of 0 is made.

We now discuss the power spectral density of the transmitted signal. The transmitted BPSK signal can be expressed as

$$s(t) = \sqrt{E_b} \sum_{k=-\infty}^{\infty} b_k \phi_1(t - kT_b), \tag{5-33}$$

where $b_k \in \{+1, -1\}$. Note that $\phi_1(t)$ is actually a truncated cosine waveform as shown in Figure 5-11.

Let $h(t) = \sqrt{E_b}\phi_1(t)$. The power spectral density for $s(t)$ can be calculated from Equation (5-10):

$$P_s(f) = \frac{1}{T_b}|H(f)|^2. \tag{5-34}$$

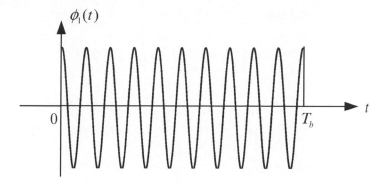

Figure 5-11   A truncated cosine waveform $\phi_1(t)$

The problem now is to find the Fourier transform of $h(t)$. The function $\phi_1(t)$ can be expressed as

$$\phi_1(t) = \sqrt{\frac{2}{T_b}}\Pi\left(\frac{t - T_b/2}{T_b}\right)\cos(2\pi f_c t), \qquad (5\text{-}35)$$

where $\Pi(x)$ is called the *rectangular function* and is defined as follows:

$$\Pi(x) = \begin{cases} 1 & \text{if } -\dfrac{1}{2} \leq x \leq \dfrac{1}{2} \\ 0 & \text{otherwise.} \end{cases} \qquad (5\text{-}36)$$

It can be easily seen that

$$\Pi\left(\frac{t - T_b/2}{T_b}\right) = 1 \quad \text{for} \quad 0 \leq t \leq T_b$$

$$\Pi\left(\frac{t - T_b/2}{T_b}\right) = 0 \quad \text{otherwise.}$$

This is why $\phi_1(t)$ is a product of $\Pi\{(t - T_b/2)/T_b\}$ and $\cos(2\pi f_c t)$.

Then we may use the property that the Fourier transform of the product of two functions in the time domain is equivalent to the convolution of the Fourier transforms of these two functions. This property was discussed in Chapter 3. We thus have

$$H(f) = \sqrt{\frac{2E_b}{T_b}}F\left[\Pi\left(\frac{t - T_b/2}{T_b}\right)\right] * F[\cos(2\pi f_c t)]. \qquad (5\text{-}37)$$

From Equation (3-102), the Fourier transform of the rectangular function is

$$F\left[\Pi\left(\frac{t - T_b/2}{T_b}\right)\right] = T_b\text{sinc}(T_b f)e^{-j\pi T_b f}, \qquad (5\text{-}38)$$

and from Equation (3-145) we have

$$F[\cos(2\pi f_c t)] = \frac{1}{2}\delta(f - f_c) + \frac{1}{2}\delta(f + f_c). \tag{5-39}$$

We can now see that

$$H(f) = \sqrt{\frac{E_b T_b}{2}}\mathrm{sinc}(T_b(f - f_c))e^{-j\pi T_b(f-f_c)} + \sqrt{\frac{E_b T_b}{2}}\mathrm{sinc}(T_b(f + f_c))e^{-j\pi T_b(f+f_c)}. \tag{5-40}$$

Assume that $f_c \gg 1/T_b$. Then

$$|H(f)|^2 = \frac{E_b T_b}{2}\mathrm{sinc}^2(T_b(f - f_c)) + \frac{E_b T_b}{2}\mathrm{sinc}^2(T_b(f + f_c)). \tag{5-41}$$

Thus the power spectral density is

$$\begin{aligned} P_s(f) &= \frac{1}{T_b}|H(f)|^2 \\ &= \frac{E_b}{2}\mathrm{sinc}^2(T_b(f - f_c)) + \frac{E_b}{2}\mathrm{sinc}^2(T_b(f + f_c)). \end{aligned} \tag{5-42}$$

The power spectral density for the case with $E_b = 1$, $T_b = 1\,\mu s$ and $f_c = 6\,\text{MHz}$ is illustrated in Figure 5-12. The data rate for this system is $R = 1/T_b = 1\,\text{MBit/s}$. The main power of the signal concentrates over the carrier frequency $f_c = 6\,\text{MHz}$. The bandwidth of the signal is 2 MHz.

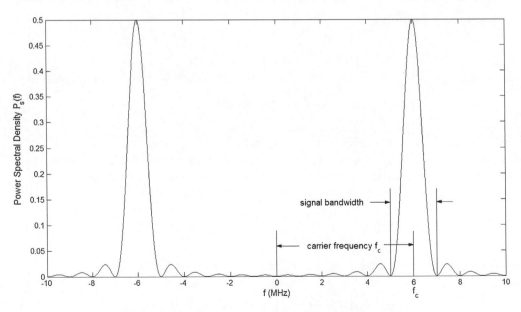

Figure 5-12   Power spectral density for a BPSK signal

The above discussion is valid not only for the BPSK case, it is also valid for the ASK case. We can see that, unlike with baseband pulse transmission, we now use a carrier frequency to transmit the signal, as we do in the ASK case. The signal bandwidth $2/T_b$ is determined by the data rate, as in the baseband pulse transmission system. The higher the data rate, the smaller T_b and the larger $2/T_b$. Suppose that the bit rate is 1 Mbits/s. The bandwidth of the transmitted signal will then be 2 MHz.

5.4 Binary Frequency-shift Keying (FSK)

In a binary FSK system, symbols 1 and 0 are distinguished from each other by transmitting one of two sinusoidal waves that differ in frequency by a fixed amount. A typical pair of sinusoidal waves is

$$s_1(t) = \sqrt{\frac{2E_b}{T_b}} \cos(2\pi f_1 t)$$

$$s_2(t) = \sqrt{\frac{2E_b}{T_b}} \cos(2\pi f_2 t)$$

(5-43)

for $0 \leq t < T_b$, where E_b is the transmitted signal energy per bit. The transmitted frequency is

$$f_i = \frac{n_c + i}{2T_b} \quad \text{for some fixed integer } n_c, i = 1, 2.$$

(5-44)

Note that $f_2 - f_1 = 1/(2T_b)$ and symbol 1 is represented by $s_1(t)$ and symbol 0 by $s_2(t)$. It is easy to show that $s_1(t)$ and $s_2(t)$ are orthogonal. In fact, it can be shown that $1/(2T_b)$ is the minimum frequency spacing $f_2 - f_1$ between two signals to be orthogonal. Define:

$$\phi_i(t) = \sqrt{\frac{2}{T_b}} \cos(2\pi f_i t), \qquad i = 1, 2, 0 \leq t < T_b.$$

(5-45)

We will show that $\phi_1(t)$ and $\phi_2(t)$ are orthonormal basis functions.

First, we want to show that $\phi_1(t)$ and $\phi_2(t)$ have unit energy. The energy of the signal $\phi_i(t)$ over $(0,T_b)$ is given by

$$\int_0^{T_b} \phi_i^2(t)dt = \frac{2}{T_b} \int_0^{T_b} \cos^2(2\pi f_i t)dt$$

$$= \frac{2}{T_b} \left(\int_0^{T_b} \frac{1}{2}dt + \int_0^{T_b} \frac{1}{2}\cos(4\pi f_i t)dt \right)$$

(5-46)

$$= \frac{2}{T_b} \left(\frac{t}{2} \Big|_0^{T_b} + \frac{1}{8\pi f_i}\sin(4\pi f_i t) \Big|_0^{T_b} \right).$$

Substituting $f_i = (n_c + i)/(2T_b)$ into the above equation, we have

$$\int_0^{T_b} \phi_i^2(t)dt = \frac{2}{T_b}\left(\left(\frac{T_b}{2} - 0\right) + \frac{1}{8\pi(n_c + i)/(2T_b)}\left(\sin\left(4\pi\frac{n_c + i}{2T_b}T_b\right) - 0\right)\right)$$

$$= \frac{2}{T_b}\left(\frac{T_b}{2} + \frac{1}{8\pi(n_c + i)/(2T_b)}\sin(2(n_c + i)\pi)\right) \tag{5-47}$$

$$= 1.$$

The last equality follows since $n_c + i$ is a positive integer and hence the second term in the brackets is zero. In addition:

$$\int_0^{T_b} \phi_1(t)\phi_2(t)dt = \frac{2}{T_b}\int_0^{T_b} \cos(2\pi f_1 t)\cos(2\pi f_2 t)dt$$

$$= \frac{1}{T_b}\int_0^{T_b} \cos(2\pi(f_1 + f_2)t) + \cos(2\pi(f_1 - f_2)t)dt. \tag{5-48}$$

Substituting $f_i = (n_c + i)/(2T_b)$ into the above equation, we have

$$\int_0^{T_b} \phi_1(t)\phi_2(t)dt = \frac{1}{T_b}\left(\int_0^{T_b} \cos\left(2\pi\frac{(2n_c + 3)t}{2T_b}\right)dt + \int_0^{T_b} \cos\left(2\pi\frac{1}{2T_b}t\right)dt\right)$$

$$= \frac{1}{T_b}\left(\int_0^{T_b} \cos\left(\pi\frac{(2n_c + 3)t}{T_b}\right)dt + \int_0^{T_b} \cos\left(\pi\frac{1}{T_b}t\right)dt\right) \tag{5-49}$$

We may show that, if n is a positive integer, the following holds:

$$\int_0^{T_b} \cos\left(\pi\frac{n}{T_b}t\right)dt = \frac{T_b}{\pi n}\sin\left(\pi\frac{n}{T_b}t\right)\Big|_0^{T_b} = \frac{T_b}{\pi n}\sin(\pi n) - \frac{T_b}{\pi n}\sin(0) = 0.$$

Therefore, the first term of Equation (5-49) corresponds to $n = 2n_c + 3$ and hence the first term of (5-49) is zero. Similarly, the second term of (5-49) is also zero. As a result, we have

$$\int_0^{T_b} \phi_1(t)\phi_2(t)dt = 0. \tag{5-50}$$

Therefore, $\phi_1(t)$ and $\phi_2(t)$ form an orthonormal basis.

There is a simpler way to prove that $\phi_1(t)$ and $\phi_2(t)$ are orthonormal. Let $f_0 = 1/(2T_b)$. Then, by Equation (5-44), we have $f_2 = (n_c + 2)f_0$ and $f_1 = (n_c + 1)f_0$. By using Equations (3-6) and (3-7), we can prove that $\phi_1(t)$ and $\phi_2(t)$ are orthonormal.

To generate a binary FSK signal, we may use the scheme shown in Figure 5-13. The binary data sequence is first applied to an on–off level encoder. At the output of the encoder, symbol 1 is represented by a constant amplitude of $\sqrt{E_b}$ and symbol 0 is represented by zero. An inverter is used in the lower channel. If the input of the inverter is 0, the output is

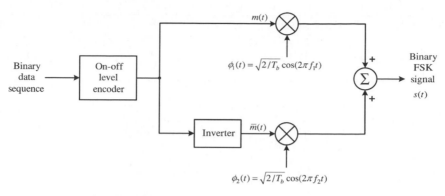

Figure 5-13 Modulator for an FSK system

$\sqrt{E_b}$. On the other hand, if the input is $\sqrt{E_b}$, the output will be 0. The multipliers act as switches which turn the corresponding carrier on and off. If the input of the multiplier is $\sqrt{E_b}$ (0), the corresponding carrier will be turned on (off). Now, if the transmitted symbol is 1, the carrier of the upper channel will be turned on, and that of the lower channel will be turned off. Similarly, if the transmitted symbol is 0, the carrier of the upper channel will be turned off, and that of the lower channel will be turned on. Therefore, the resulting output signal is altered between two carriers of different frequencies controlled by the input symbol.

Figure 5-14 gives an example of resulting waveform when the input binary stream is 1001101, where $E_b = 1$, $T_b = 1$ s, $f_1 = 3$ Hz and $f_2 = 3.5$ Hz.

Demodulation of the received signals utilizes the fact that $\phi_1(t)$ and $\phi_2(t)$ are orthonormal. That is, if we want to detect $\phi_1(t)$, we multiply the received signal by $\phi_1(t)$ and integrate. Because $\phi_1(t)$ and $\phi_2(t)$ are orthonormal, the $\phi_2(t)$ part will disappear totally and only the coefficient of $\phi_1(t)$ remains. We may also multiply the received signal by $\phi_2(t)$ and integrate. This time, only the coefficient related to $\phi_2(t)$ remains. Since only one of the coefficients is nonzero, we can easily see whether 1 or 0 is transmitted.

The above discussion leads to the receiver shown in Figure 5-15. It consists of two correlators with a common input, which are supplied with two locally generated coherent signals $\phi_1(t)$ and $\phi_2(t)$. The correlator outputs are then subtracted, one from the other, and the resulting difference, y, is compared with a threshold of zero. If $y > 0$, the receiver decides in favor of 1. If $y < 0$, it decides in favor of 0.

For the noiseless case, the received signal is exactly the transmitted signal. If the signal $s_1(t)$ (representing 1) was transmitted, the outputs x_1 and x_2 of the correlators can be expressed as

$$x_1 = \int_0^{T_b} \sqrt{\frac{2E_b}{T_b}} \cos(2\pi f_1 t) \sqrt{\frac{2}{T_b}} \cos(2\pi f_1 t) dt = \sqrt{E_b}$$

$$x_2 = \int_0^{T_b} \sqrt{\frac{2E_b}{T_b}} \cos(2\pi f_1 t) \sqrt{\frac{2}{T_b}} \cos(2\pi f_2 t) dt = 0.$$

(5-51)

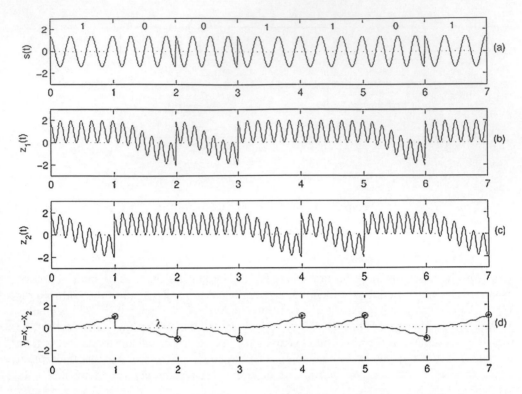

Figure 5-14 Signals for an FSK system: (a) transmitted signal; (b) multiplication output $z_1(t) = s(t)\sqrt{2/T_b}\cos(2\pi f_1 t)$; (c) multiplication output $z_2(t) = s(t)\sqrt{2/T_b}\cos(2\pi f_2 t)$; (d) difference between two outputs of the integrators and the corresponding sampling points

On the other hand, if the signal $s_2(t)$ (representing 0) was transmitted, the outputs of the correlators can be expressed as

$$x_1 = \int_0^{T_b} \sqrt{\frac{2E_b}{T_b}}\cos(2\pi f_2 t)\sqrt{\frac{2}{T_b}}\cos(2\pi f_1 t)dt = 0$$

$$x_2 = \int_0^{T_b} \sqrt{\frac{2E_b}{T_b}}\cos(2\pi f_2 t)\sqrt{\frac{2}{T_b}}\cos(2\pi f_2 t)dt = \sqrt{E_b}.$$

(5-52)

Therefore, $y = x_1 - x_2$ is greater than zero if $s_1(t)$ was transmitted and is less than zero if $s_2(t)$ was transmitted.

Figure 5-14 illustrates the signals at the receiver. The transmitted signal is given in 5-14(a). The received signal is first multiplied by two carriers with frequencies 3 and 3.5 Hz, respectively: the results are shown in 5-14(b) and 5-14(c), respectively. Finally, the resulting difference $y = x_1 - x_2$ is given in 5-14(d). The decision device based on the sampled values makes a decision of 1001101 which is exactly the same as the transmitted sequence.

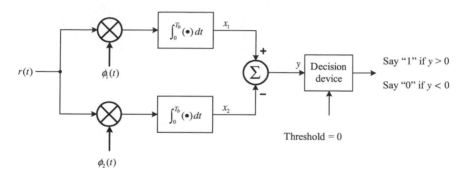

Figure 5-15   Coherent detector for an FSK system

Perhaps it is helpful to think along the following lines. In the FSK system, we are using two functions, say $\phi_1(t)$ and $\phi_2(t)$, to represent one bit. If the bit is 1 (or 0), we send $\phi_1(t)$ (or $\phi_2(t)$) under the condition that $\phi_1(t)$ and $\phi_2(t)$ are orthornormal to each other. For each bit, the sent signal is represented by

$$s(t) = a_1\phi_1(t) + a_2\phi_2(t).$$

At the receiver, demodulation involves performance of the inner product of $s(t)$ and $\phi_i(t)$ for $i = 1,2$:

$$\langle s(t), \phi_1(t)\rangle = a_1\langle\phi_1(t), \phi_1(t)\rangle + a_2\langle\phi_2(t), \phi_1(t)\rangle = a_1$$

$$\langle s(t),\ \phi_2(t)\rangle = a_1\langle\phi_1(t), \phi_2(t)\rangle + a_2\langle\phi_2(t), \phi_2(t)\rangle = a_2.$$

Thus, after performing these inner products, we will find the values of a_1 and a_2 and determine whether 1 or 0 was sent.

We have shown that there are two orthonormal basis functions $\phi_1(t)$ and $\phi_2(t)$ defined in Equation (5-45). The signal points $s_1(t)$ and $s_2(t)$ can be expressed as

$$s_i(t) = s_{i1}\phi_1(t) + s_{i2}\phi_2(t), i = 1,2, \quad 0 \leq t < T_b. \tag{5-53}$$

Therefore a coherent binary FSK system is characterized by having a signal space that is two-dimensional with two message points, as shown in Figure 5-16. The two message points are defined by $s_1 = (\sqrt{E_b}, 0)$ and $s_2 = (0, \sqrt{E_b})$. As shown in Figure 5-16, the decision boundary is characterized by the line with $x_1 - x_2 = 0$. Therefore, if $x_1 - x_2 > 0$, the received signal point lies in region Z_1 and the decision device makes a decision of 1. If $x_1 - x_2 < 0$, the received signal point lies in region Z_2 and the decision device makes a decision of 0.

The FSK system, like the ASK system, has a noncoherent demodulation method. We only need to implement two band-pass filters followed by envelope detectors. The first band-pass filter passes the signal with frequency f_1 and the second one passes the signal with frequency

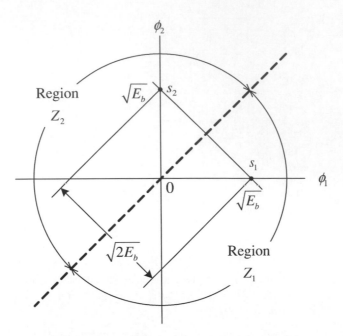

Figure 5-16   Signal space diagram for binary FSK

f_2. The filtered signals are then passed separately through the two envelope detectors for power detection. For example, if a symbol 1 is transmitted, the received signal will concentrate over the frequency of f_1. In this case, the signal will pass through the first band-pass filter and will be rejected by the second band-pass filter. Therefore, a strong signal appears at the output of the first band-pass filer and a weak signal appears at the output of the second band-pass filter. If we compare the powers of both outputs, we may find that symbol 1 was actually transmitted.

The distances between two signal points for the binary ASK, BPSK and FSK systems are $\sqrt{2E_b}$, $2\sqrt{E_b}$ and $\sqrt{2E_b}$, respectively, as shown in Figures 5-7, 5-10 and 5-16. Therefore, for the same transmitted bit energy and the same level of noise corruption, the BPSK system gives the lowest bit error rate since the BPSK system has the largest distance between two signal points. The performances of the ASK and FSK systems are exactly the same in this respect since they have the same distance between two signal points.

It can be calculated that the bandwidth of the FSK system is approximately $2/T_b$. There are two carrier frequencies f_1 and f_2 now, depending on whether we are sending 1 or 0.

The FSK system, like the ASK system, has a noncoherent demodulation method. We need to implement only two band-pass filters. One passes the signal with frequency f_1 and the other one passes the signal with frequency f_2. In this way, we can determine whether the sent signal is 1 or 0.

Many house automation and remote control systems use FSK transceivers. In a typical transceiver the frequencies are 433.93 and 434.33 MHz, the maximum bit rate is 20 kBit/s and the bandwidth is 30 kHz.

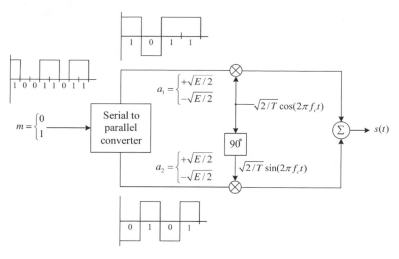

Figure 5-17   Schematic diagram of QPSK

5.5 Quadriphase-shift Keying (QPSK)

One important goal in the design of a digital communications system is to achieve a very low probability of errors occurring. Another important goal is the efficient utilization of channel bandwidth. In this section, we study a bandwidth-conservation modulation scheme known as *coherent quadriphase-shift keying.* **Unlike with binary ASK, BPSK and FSK systems, two bits are transmitted simultaneously during one signaling interval T.** Without increasing the transmitted bandwidth, we can double the transmitted bit rate. This will be made clear later.

The QPSK system is illustrated in Figure 5-17. Let us assume that there are two bits which are going to be transmitted during one signaling interval T. They are denoted as m_1 and m_2. The pair of m_1 and m_2 is separated from a single bit stream m; that is, m_1 represents the odd bits of m and m_2 represents the even bits. As shown, m_1 **will go up and** m_2 **will go down.** Then the following rules will follow:

1. m_1 will trigger signal a_1 and m_2 will trigger signal a_2.
2. If m_1 is equal to 1 (or 0), a_1 is set to be $+\sqrt{E/2}$ (or $-\sqrt{E/2}$).
3. If m_2 is equal to 1 (or 0), a_2 is set to be $+\sqrt{E/2}$ (or $-\sqrt{E/2}$).
4. $a_1(a_2)$ will be multiplied by $\sqrt{2/T}\cos(2\pi f_c t)$ (or $\sqrt{2/T}\sin(2\pi f_c t)$).
5. The signal transmitted at any time t is

$$s(t) = a_1\sqrt{\frac{2}{T}}\cos(2\pi f_c t) + a_2\sqrt{\frac{2}{T}}\sin(2\pi f_c t). \tag{5-54}$$

The relationship between (m_1, m_2) and (a_1, a_2) is as shown in Table 5-1.

Table 5-1 The mapping of (m_1, m_2) to (a_1, a_2)

(m_1, m_2)	(a_1, a_2)
(1,1)	$(\sqrt{E}/2, \sqrt{E}/2)$
(1,0)	$(\sqrt{E}/2, -\sqrt{E}/2)$
(0,1)	$(-\sqrt{E}/2, \sqrt{E}/2)$
(0,0)	$(-\sqrt{E}/2, -\sqrt{E}/2)$

Before introducing the demodulation process of QPSK, we shall first introduce an interesting property of the system. Let us take a look at $s(t)$ in Equation (5-54). We can easily show that

$$s(t) = \sqrt{\frac{2}{T}} r \cos(2\pi f_c t - \theta), \tag{5-55}$$

where $r = \sqrt{a_1^2 + a_2^2}$ and $\theta = \tan^{-1}(a_2/a_1)$. Since $|a_1| = |a_2| = \sqrt{E/2}$, $r = \sqrt{E}$. As for θ, note that it is related to the values of a_1 and a_2. Since each of them assumes two values, there are four possible combinations. Therefore there are four possible values of θ, corresponding to four distinct combinations of a_1 and a_2. Every θ assumes an integer multiplication of $\pi/4$.

From Equation 5-55, we can see that there are four different $s(t)$, denoted as $s_i(t)$ for $i = 1, 2, 3, 4$:

$$s_i(t) = \sqrt{\frac{2E}{T}} \cos\left(2\pi f_c t + (2i - 1)\frac{\pi}{4}\right) \quad \text{for} \quad 0 \le t < T. \tag{5-56}$$

Consider the case where $m_1 = 1$ and $m_2 = 0$. In this case, $a_1 = \sqrt{E/2}$ and $a_2 = \sqrt{E/2}$. Thus we have

$$
\begin{aligned}
s_1(t) &= \sqrt{\frac{E}{T}}(\cos(2\pi f_c t) - \sin(2\pi f_c t)) \\
&= \sqrt{\frac{E}{T}}\left(\sqrt{2}\left(\frac{\sqrt{2}}{2}\cos(2\pi f_c t) - \frac{\sqrt{2}}{2}\sin(2\pi f_c t)\right)\right) \\
&= \sqrt{\frac{2E}{T}}\left(\cos\left(\frac{\pi}{4}\right)\cos(2\pi f_c t) - \sin\left(\frac{\pi}{4}\right)\sin(2\pi f_c t)\right) \\
&= \sqrt{\frac{2E}{T}}\cos\left(2\pi f_c t + \frac{\pi}{4}\right).
\end{aligned}
\tag{5-57}
$$

It can be noted that Equation (5-57) corresponds to the case where $i = 1$. Let us expand Equation (5-56) as follows:

$$
\begin{aligned}
s_i(t) &= \sqrt{\frac{2E}{T}}\cos\left((2i - 1)\frac{\pi}{4}\right)\cos(2\pi f_c t) \\
&\quad - \sqrt{\frac{2E}{T}}\sin\left((2i - 1)\frac{\pi}{4}\right)\sin(2\pi f_c t).
\end{aligned}
\tag{5-58}
$$

Table 5-2 Signal-space characterization of QPSK

	Input digit (m_1m_2)	Phase	a_1	a_2	$s_i(t)$
$s_1(t)$	10	$\pi/4$	$+\sqrt{E/2}$	$-\sqrt{E/2}$	$\sqrt{2E/T}\cos(2\pi f_c t + \pi/4)$
$s_2(t)$	00	$3\pi/4$	$-\sqrt{E/2}$	$-\sqrt{E/2}$	$\sqrt{2E/T}\cos(2\pi f_c t + 3\pi/4)$
$s_3(t)$	01	$5\pi/4$	$-\sqrt{E/2}$	$+\sqrt{E/2}$	$\sqrt{2E/T}\cos(2\pi f_c t + 5\pi/4)$
$s_4(t)$	11	$7\pi/4$	$+\sqrt{E/2}$	$+\sqrt{E/2}$	$\sqrt{2E/T}\cos(2\pi f_c t + 7\pi/4)$

We can now see that

$$a_1 = +\sqrt{E}\,\cos\left((2i-1)\frac{\pi}{4}\right) \quad \text{and} \quad a_2 = -\sqrt{E}\,\sin\left((2i-1)\frac{\pi}{4}\right). \tag{5-59}$$

It can be easily seen that $s_i(t)$, for $i = 1, 2, 3, 4$, corresponds to $m_1m_2 = 10, 00, 01, 11$, respectively. This is summarized in Table 5-2. An example of the four signals is given in Figure 5-18. The reader may easily see that each signal is a cosine function with a phase shift. This is due to Equation (5-56).

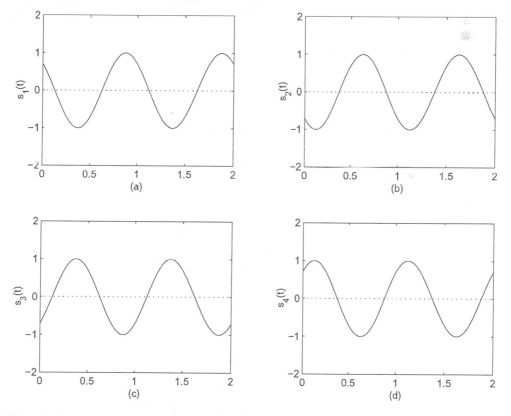

Figure 5-18 The four QPSK signals $s_i(t)$ for $i = 1, 2, 3, 4$, where $f_c = 1$: (a) signal of $s_1(t)$; (b) signal of $s_2(t)$; (c) signal of $s_3(t)$; (d) signal of $s_4(t)$

Let us now illustrate the QPSK system by using the input data stream in Figure 5-17. The stream of input data is 10011011. The odd-numbered bits go up and the even-numbered bits go down. Thus the upper stream is 1011 and the lower stream is 0101. At each time slot, one bit from the upper stream is combined with a corresponding bit from the lower stream. Thus, the output data stream is (10, 01, 10, 11). From Table 5-2, we can see that the signals sent for these four time slots are $s_1(t)$, $s_3(t)$, $s_1(t)$ and $s_4(t)$.

We now discuss how to demodulate the received signal. Note that the signal $s_i(t)$ is determined by m_1 and m_2. We shall show later that after this signal is received, it will be demodulated to such an extent that m_1 and m_2 will be correctly identified. Let us consider Equation (5-54) and rewrite it as follows:

$$s(t) = A\cos(2\pi f_c t) + B\sin(2\pi f_c t).$$

The job of demodulation is to detect A and B. Since $\cos(2\pi f_c t)$ and $\sin(2\pi f_c t)$ are orthogonal, to detect A we multiply $s_i(t)$ by $\cos(2\pi f_c t)$ and integrate. In this way, we eliminate B and only A remains. Similarly, to detect B we multiply $s_i(t)$ by $\sin(2\pi f_c t)$ and integrate. Now, A disappears and only B remains. This is the basic principle of QPSK demodulation.

Based on the principle presented in the above paragraph, demodulation at the receiver can be accomplished by coherent demodulation with two reference sinusoids that are ideally phase- and frequency-coherent with the carriers in the transmitter. A QPSK detector is shown in Figure 5-19.

The received signal is first multiplied by two sinusoids $\sqrt{2/T}\cos(2\pi f_c t)$ and $\sqrt{2/T}\sin(2\pi f_c t)$. The results are then integrated respectively by two integrators. The

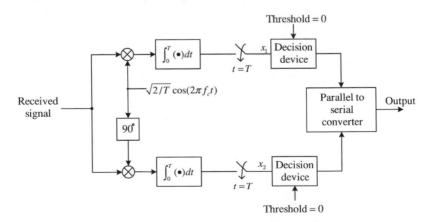

Figure 5-19   Demodulator for QPSK

outputs of the integrators are sampled at $t = T$. For the noiseless case, the resulting sampled output x_1 can be expressed as

$$x_1 = \int_0^T \left(a_1 \sqrt{\frac{2}{T}} \cos(2\pi f_c t) + a_2 \sqrt{\frac{2}{T}} \sin(2\pi f_c t) \right) \times \sqrt{\frac{2}{T}} \cos(2\pi f_c t) dt$$

$$= a_1 \frac{2}{T} \left(\int_0^T \cos^2(2\pi f_c t) dt \right) + a_2 \frac{2}{T} \left(\int_0^T \sin(2\pi f_c t) \cos(2\pi f_c t) dt \right) \qquad (5\text{-}60)$$

$$= a_1 \frac{2}{T} \left(\int_0^T \frac{1}{2} + \frac{1}{2} \cos(4\pi f_c t) dt \right) + a_2 \frac{2}{T} \left(\int_0^T \frac{1}{2} \sin(4\pi f_c t) dt \right).$$

Replacing $f_c = n_c/T$ in the above equation, where n_c is a positive integer, we obtain

$$x_1 = a_1 \frac{2}{T} \left(\frac{t}{2} + \frac{1}{8\pi(n_c)/T} \sin\left(4\pi \frac{n_c}{T} t\right) \right) \Big|_0^T + a_2 \frac{2}{T} \left(-\frac{1}{8\pi(n_c)/T} \cos\left(4\pi \frac{n_c}{T} t\right) \right) \Big|_0^T.$$

$$= a_1 \frac{2}{T} \left(\left(\frac{T}{2} + 0 \right) - (0 + 0) \right) + a_2 \frac{2}{T} \left(-\frac{1}{8\pi(n_c/T)} + \frac{1}{8\pi(n_c/T)} \right). \qquad (5\text{-}61)$$

The last equality follows since $\sin(4\pi n_c) = 0$ and $\cos(4\pi n_c) = 1$ for any positive integer n_c. Therefore, from the above equation, we have

$$x_1 = a_1. \qquad (5\text{-}62)$$

Similarly, we can show that

$$x_2 = \int_0^T \left(a_1 \sqrt{\frac{2}{T}} \cos(2\pi f_c t) + a_2 \sqrt{\frac{2}{T}} \sin(2\pi f_c t) \right) \times \sqrt{2/T} \sin(2\pi f_c t) dt = a_2. \qquad (5\text{-}63)$$

Therefore, the original messages a_1 and a_2 can be separated at the receiver and hence can be detected independently. The sampled outputs x_1 and x_2 are each compared with a threshold of zero by the decision devices located at the upper and lower channels, respectively. If $x_1 > 0$, a decision is made in favor of $m_1 = 1$ for the upper channel; but if $x_1 < 0$ a decision is made in favor of $m_1 = 0$. Similarly, for the lower channel, if $x_2 > 0$, a decision is made in favor of $m_2 = 1$; but if $x_2 < 0$ a decision is made in favor of $m_2 = 0$. Finally, these two binary sequences are combined in a parallel-to-serial converter to reproduce the original binary sequence at the transmitter input.

An example of QPSK signals is given in Figure 5-20. In this case, $f_c = 1$. The transmitted signal $s(t)$ is given in Figure 5-20(a) which is a modulated signal with the input bit stream 10011011, where $E = 1$, $f_c = 1$ Hz and $T = 2$ s. At the receiver, the signals after the multiplication of carriers denoted by $y_1(t) = s(t) \times \sqrt{2/T} \cos(2\pi f_c t)$ and $y_2(t) = s(t) \times \sqrt{2/T} \sin(2\pi f_c t)$ are given in 5-20(b) and 5-20(c), respectively. The outputs from the integrators and the corresponding sampling points x_1 and x_2 are in 5-20(d) and 5-20(e), respectively. Based on the sampled values, the decision output after the parallel-to-serial converter is 1001101, which is exactly the same as the transmitted sequence.

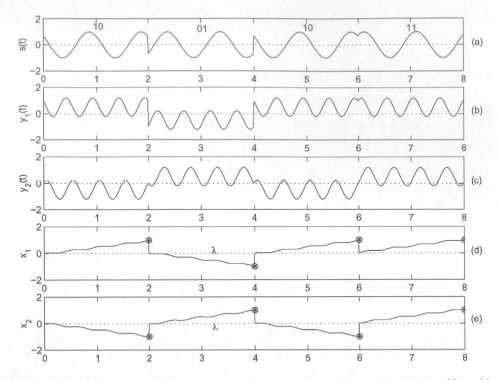

Figure 5-20 Signals of QPSK for $f_c = 1$: (a) transmitted signal; (b) signal $y_1(t) = s(t) \times \sqrt{2/T}\cos(2\pi f_c t)$; (c) signal $y_2(t) = s(t) \times \sqrt{2/T}\sin(2\pi f_c t)$; (d) output of the upper integrator and the corresponding sampling points; (e) output of the lower integrator and the corresponding sampling points

Perhaps it will be interesting to examine the function $y_1(t)$. Consider the case where the input datum is 10. According to Table 5-2:

$$s(t) = \sqrt{\frac{2E}{T}}\cos\left(2\pi t + \frac{\pi}{4}\right).$$

Thus:

$$y_1(t) = \sqrt{\frac{2E}{T}}\cos\left(2\pi t + \frac{\pi}{4}\right)\sqrt{\frac{2}{T}}\cos(2\pi t)$$

$$= \frac{2\sqrt{E}}{T}\left(\frac{1}{2}\right)\left(\cos\left(4\pi t + \frac{\pi}{4}\right) + \cos\left(\frac{\pi}{4}\right)\right)$$

$$= \frac{\sqrt{E}}{T}\left(\cos\left(4\pi t + \frac{\pi}{4}\right) + \cos\left(\frac{\pi}{4}\right)\right)$$

$$= \frac{\sqrt{E}}{T} \left(2\cos^2\left(2\pi t + \frac{\pi}{8}\right) - 1 + \frac{\sqrt{2}}{2} \right)$$

$$= \frac{\sqrt{E}}{T} \left(2\cos^2\left(2\pi t + \frac{\pi}{8}\right) - 1 + 0.707 \right)$$

$$= \frac{\sqrt{E}}{T} \left(2\cos^2\left(2\pi t + \frac{\pi}{8}\right) - 0.293 \right).$$

Thus $y_1(t)$, for this case, is positive for most of the time, as shown in Figure 5-20(b).

Let us here emphasize the basic principle of the QPSK system. *The QPSK system mixes two bits together and transmits them at the same time. Why can they be detected correctly? This is due to the fact that there are two orthonormal functions $\phi_1(t)$ and $\phi_2(t)$ contained in the expansion of $s_i(t)$. Specifically, $\phi_1(t)$ and $\phi_2(t)$ are given by*

$$\phi_1(t) = \sqrt{\frac{2}{T}}\cos(2\pi f_c t) \quad 0 \le t \le T$$

$$\phi_2(t) = \sqrt{\frac{2}{T}}\sin(2\pi f_c t) \quad 0 \le t \le T. \tag{5-64}$$

It is easy to verify that $\phi_1(t)$ and $\phi_2(t)$ are orthonormal basis functions; that is:

$$\int_0^T \phi_1(t)\phi_2(t)dt = 0 \tag{5-65}$$

$$\int_0^T \phi_i^2(t)dt = 1 \quad \text{for } i = 1, 2. \tag{5-66}$$

The transmitted signal therefore can be expressed as

$$s_i(t) = s_{i1}\phi_1(t) + s_{i2}\phi_2(t). \tag{5-67}$$

A look at Equation (5-67) will tell us that the demodulation of the QPSK system can be done by multiplying $s_i(t)$ by $\phi_1(t)$ and $\phi_2(t)$ and integrating over [0,T]. Multiplying $s_i(t)$ by $\phi_1(t)$ and integrating over [0,T] will give us the value of s_{i1} and multiplying $s_i(t)$ by $\phi_2(t)$ and integrating over [0,T] will give us the value of s_{i2}. Once s_{i1} and s_{i2} are known, m_1 and m_2 can be determined accordingly.

In the QPSK system, there are four vectors $s_i = (s_{i1}, s_{i2})$ for $i = 1, 2, 3, 4$ which can be represented as signal points in the signal-space diagram as shown in Figure 5-21.

It is possible to increase the transmitted bit rate by increasing the number of signals in the signal constellation. For example, we may double the number of QPSK signals, resulting in 8 signal points as shown in Figure 5-22. This signal constellation is called 8PSK which can transmit three bits at each signaling interval without increasing the transmitted bandwidth. Since three, instead of two, bits are now sent at the same time, the 8PSK system is obviously more efficient than the QPSK system. However, under the same transmitted signal energy per symbol, the distance between two nearest points in the 8PSK is less than that of the QPSK, so it is more likely that an 8PSK detector will make more errors. Therefore, we pay

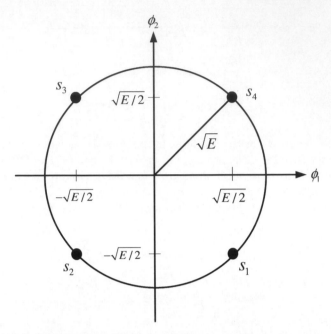

Figure 5-21   Signal-space diagram of a QPSK system

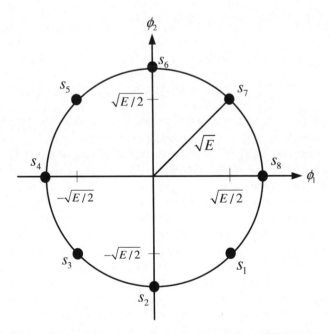

Figure 5-22   Signal-space diagram of a coherent 8PSK system

for this by sacrificing the bit error rate. To achieve the same bit error rate as that of a QPSK system, we have to increase the transmitted signal energy.

The reader may wonder how the three bits can be sent together under the 8PSK system. This will be explained in the next section.

For this QPSK system, there is a carrier frequency as the transmitted signal is a cosine function with frequency f_c. The bandwidth of the system is $2/T$, where T is the symbol duration time. For QPSK, $T = 2T_b$, so the bandwidth is $1/T_b$. If the bit rate is kept the same, T_b will remain the same, then the bandwidth is reduced to half of the case where one bit is sent at a time. This means that we may use a cheaper system because the bandwidth is half of the original. On the other hand, suppose we double the bit rate and thus reduce the bit length to half of it. Let the new bit length be denoted as $T_b' = (1/2)T_b$. Thus the symbol duration $T = 2T_b' = 2(1/2)T_b = T_b$. The QPSK bandwidth becomes $2/T = 2/T_b$. This means we can double the data rate without increasing the bandwidth. Note that normally, an increase of the data rate will always result in an increase of the bandwidth. This is a clear advantage of using the QPSK system.

5.6 Quadrature Amplitude Modulation

In the foregoing sections we have introduced BPSK, QPSK and 8PSK systems. Let us now gather their signal-space diagrams which were illustrated in Figures 5-10, 5-21 and 5-22 and display them again in Figure 5-23.

From this we may conclude that the BPSK, QPSK and 8PSK systems can be viewed as special cases of a more general class of digital modulation system, called the PSK (phase-shifting keying) system. The signals sent are only different in their phases. For BPSK, the phase difference is π, for QPSK it is $\pi/2$, and for 8PSK it is $\pi/4$. Thus we may generalize the concept into the *M-ary PSK system*.

Let us assume that we would like to send k bits together. We let $M = 2^k$. In general, for an M-ary PSK system, the transmitted signal can be represented by

$$s_i(t) = \sqrt{\frac{2E}{T}} \cos\left(2\pi f_c t + (2i-1)\frac{\pi}{M}\right) \quad \text{for} \quad i = 1, 2, \ldots, M. \tag{5-68}$$

Each $s_i(t)$ represents a possible state of sending k bits together. Similar to Equation (5-67), the signal can be equivalently written by

$$s_i(t) = s_{i1}\phi_1(t) + s_{i2}\phi_2(t) \quad \text{for} \quad i = 1, 2, \ldots, M, \tag{5-69}$$

where $\phi_1(t)$ and $\phi_2(t)$ are defined in (5-64) as follows:

$$\phi_1(t) = \sqrt{\frac{2}{T}}\cos(2\pi f_c t) \qquad 0 \le t \le T$$
$$\phi_2(t) = \sqrt{\frac{2}{T}}\sin(2\pi f_c t). \qquad 0 \le t \le T \tag{5-70}$$

It can be easily seen that

$$s_{i1} = \sqrt{E}\cos\left((2i-1)\frac{\pi}{M}\right) \quad \text{and} \quad s_{i2} = -\sqrt{E}\sin\left((2i-1)\frac{\pi}{M}\right). \tag{5-71}$$

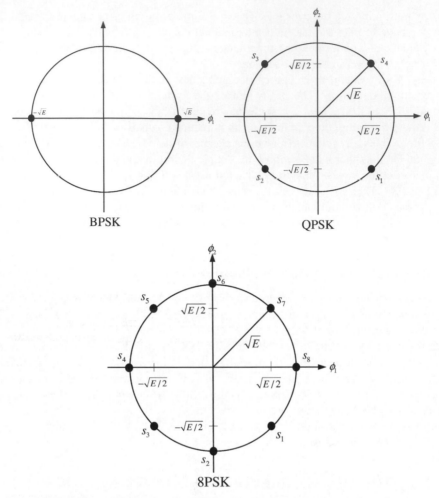

Figure 5-23   Signal-space diagrams for BPSK, QPSK and 8PSK systems

We may plot the locations of the vectors (s_{i1}, s_{i2}) on a signal-space diagram. The result shows that, for PSK systems, the locations of the vector (s_{i1}, s_{i2}) are on a circle of radius $\sqrt{E}$. For example, for the QPSK modulation, the four vectors $s_i = (s_{i1}, s_{i2})$ for $i = 1, 2, 3, 4$ are on a circle as shown in Figure 5-21. For the 8PSK modulation, the eight vectors $s_i = (s_{i1}, s_{i2})$ for $i = 1, 2, \cdots, 8$ are on a circle as shown in Figure 5-22. Therefore, in an M-ary PSK system, the message points are always on a circle of radius $\sqrt{E}$.

The QPSK system was illustrated in Figure 5-17. In this system two bits are sent together. One bit goes up and is multiplied by a cosine function while the other bit goes down and is multiplied by a sine function. For the 8PSK system, we cannot illustrate it by such a simple-minded scheme. Instead, we should think of the system as having a mapping system. In this system, three bits are sent together. Thus there are $2^3 = 8$ possible states. We may label these

Table 5-3 A possible labeling of (m_1, m_2, m_3)

(m_1, m_2, m_3)	i
(0,0,0)	1
(0,0,1)	2
(0,1,0)	3
(0,1,1)	4
(1,0,0)	5
(1,0,1)	6
(1,1,0)	7
(1,1,1)	8

8 states arbitrarily by $i = 1, 2, \cdots, 8$. Each state is represented by (m_1, m_2, m_3) where each $m_i = 1$ or 0. Each (m_1, m_2, m_3) corresponds to a number if we use the binary numbering system. For instance (0,0,0) corresponds to 0, and (1,0,0) corresponds to 4 and (0,1,1) corresponds to 3. Suppose a vector corresponds to j. We then label it $i = j + 1$. Note that this is not the only way to label the states. We may label them arbitrarily. After the labeling is done, we then use Equation (5-68) to send the signal. This discussion can be summarized by Table 5-3.

We would like to emphasize that the table illustrates only one possible way to label the (m_1, m_2, m_3) in 8PSK. Any arbitrary labeling can be used provided distinct (m_1, m_2, m_3) correspond to distinct i.

Thus the modulating algorithm for the 8PSK system is as follows.

- **Step 1:** For the particular transmitted bits (m_1, m_2, m_3), find its corresponding index i from the table relating the (m_1, m_2, m_3) to i.
- **Step 2:** Send the signal out according to Equation (5-68) by using the determined i.

So far as the receiver is concerned, it understands that the received signal is in the form of Equations (5-69) and (5-70). Since $\phi_1(t) = \sqrt{2/T} \cos(2\pi f_c t)$ and $\phi_2(t) = \sqrt{2/T} \sin(2\pi f_c t)$ are orthogonal, s_{i1} and s_{i2} can be obtained by the concept of an inner product. Once s_{i1} and s_{i2} are determined, the index i can be found by using Equation (5-71). After i is determined, (m_1, m_2, m_3) can be determined through Table 5-3. Thus the demodulating algorithm for the 8PSK system is as follows.

- **Step 1:** For the received signal $s_i(t)$, perform an inner product $\langle s_i(t), \cos(2\pi f_c t) \rangle$ $(\langle s_i(t), \sin(2\pi f_c t) \rangle)$ to determine s_{i1} (s_{i2}) based on Equations (5-69) and (5-70).
- **Step 2:** Use the value of s_{i1} or s_{i2} to find the index i based on Equation (5-71).
- **Step 3.** From the table relating (m_1, m_2, m_3) to i, determine the corresponding (m_1, m_2, m_3).

The above scheme can be used for any PSK system. When we send k bits together, we label each possible state $(x_1, x_2, \cdots, x_k)$ as a distinct i, $1 \leq i \leq 2^k$, and later when this state occurs we use Equation (5-71) to determine s_{i1} and s_{i2}.

Consider Figure 5-22 again, one can see that the eight signal points are all spread on a circle. As compared with the four signal points spread on a circle, these eight signal points are closer to one another. This means a higher error rate. If we send even more bits together, we will have even more points on the circle and an even higher error rate. Now we shall introduce a method to send bits together which avoids this problem, called the *M-ary quadrature amplitude modulation* (M-QAM) system.

In the M-QAM system, the constraint expressed in Equation (5-68) is removed, and the components s_{i1} and s_{i2} are modulated independently. An M-ary QAM signal is still expressed by

$$s_i(t) = s_{i1}\phi_1(t) + s_{i2}\phi_2(t) \quad \text{for} \quad i = 1, 2, \dots, M, \qquad (5\text{-}72)$$

where $\phi_1(t)$ and $\phi_2(t)$ are orthogonal. Since $\phi_1(t)$ and $\phi_2(t)$ are orthogonal, two independent messages can be modulated over the amplitudes s_{i1} and s_{i2} of $\phi_1(t)$ and $\phi_2(t)$, respectively. The s_{i1} and s_{i2} can take values from a finite set of numbers. For example, for 16-QAM modulation, s_{i1} and s_{i2} may take values from $\{-3, -1, +1, +3\}$. All the possible signal points (signal constellation) for the 16-QAM are shown in Figure 5-24. Since there are $16 = 2^4$ points in the signal constellation, the system is capable of transmitting 4 bits per signaling interval. Assume that the message bits are denoted by the vector

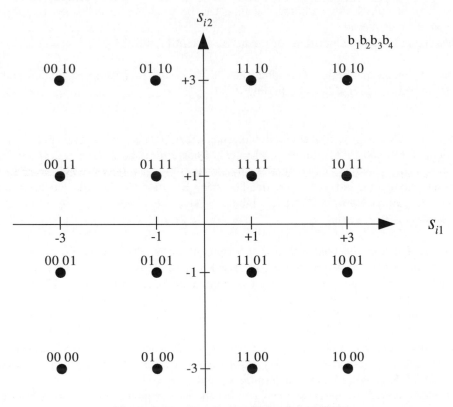

Figure 5-24 Signal constellation for 16-QAM

Table 5-4   Mapping of bit patterns (b_1, b_2, b_3, b_4) to the signal points (s_{i1}, s_{i2})

(b_1, b_2)	s_{i1}	(b_3, b_4)	s_{i2}
(0,0)	-3	(0,0)	-3
(0,1)	-1	(0,1)	-1
(1,1)	1	(1,1)	1
(1,0)	3	(1,0)	3

(b_1, b_2, b_3, b_4) with $b_i \in \{0, 1\}$. The message vector (b_1, b_2, b_3, b_4) is divided into two vectors (b_1, b_2) and (b_3, b_4). Then the first vector is mapped to s_{i1}, related to the first two bits, and the second vector is mapped to s_{i2}, related to the last two bits, according to Table 5-4.

In a similar manner, we may extend the signal constellation to 64-QAM or even higher level of signal constellation (e.g. 256-QAM). Note that the advantage of a large signal constellation is that a larger number of bits can be sent together and the transmission data rate can be increased. However, if a PSK system is used, for a transmitter with fixed transmission power, the symbol error probability is increased since the distance between signal points is decreased due to a dense constellation. Thus the M-QAM system is more desirable because the constellation is not so dense as that in a PSK system.

The advantage of using M-ary PSK and M-QAM systems is that the bandwidth will be reduced if the bit rate is kept the same and the bit rate can be increased if the bandwidth is kept the same.

5.7 Orthogonal Frequency Division Multiplexing (OFDM)

The modulation schemes introduced in earlier sections require only one carrier with a fixed carrier frequency. Now we will introduce a modulation scheme using many carriers with different frequencies. In general, such systems are called *multi-carrier systems*. *The orthogonal frequency division multiplexing (OFDM) scheme is a special case, where a single data stream is transmitted over a number of lower rate subcarriers, where every pair of subcarriers is orthogonal.*

In OFDM systems, the entire signal frequency band is divided into N frequency subchannels. Each subchannel is modulated with a separate symbol and the N signals of all the subchannels are superimposed and transmitted over the channel. Since all signals from all subchannels are mixed in the transmitter, it is important to separate the information symbols from individual subchannels at the receiver. The orthogonal property between subcarriers gives the answer to this problem. The word 'orthogonal' indicates that there is a precise mathematical relationship between the frequencies of the carriers in the system. For OFDM systems, the frequencies of all carriers are separated apart in such a way that every pair of the carrier signals is orthogonal.

An OFDM signal consists of a sum of subcarriers that are modulated by using phase-shift keying (PSK) or quadrature amplitude modulation (QAM). For simplicity, consider a binary PSK-modulated OFDM system where N carriers are employed in the system. In this case, N information bits are transmitted per OFDM symbol. *That is, in OFDM systems, N bits are transmitted in the same symbol interval.* For the ith bit, if this bit is 1 (or 0), m_i is set to be $+1$ (or -1). Let T be the duration of the OFDM symbol. Then the transmission rate is N/T bits/second. Let $(m_0, m_2, \ldots, m_{N-1})$, where $m_i \in \{\pm 1\}$ for $0 \le i \le N - 1$, be the transmitted information bits. The OFDM signal can be expressed as

$$s(t) = \sum_{k=-N/2}^{N/2-1} m_{k+N/2} \cos(2\pi(f_c + k\Delta f)t) \quad \text{for} \quad 0 < t < T, \tag{5-73}$$

where f_c is the carrier frequency and Δf is the bandwidth of each subchannel. The carrier frequency is set to be c/T with c being a positive integer greater than N. According to Section 5.3, the OFDM system can be thought as a system that consists of N BPSKs with different frequencies.

The OFDM system can be combined with the PSK or QAM system. For both PSK and QAM systems, one carrier is used to send more than one bit. In the OFDM system, there are N subcarriers. If Equation (5-73) is used, each subcarrier corresponds to one bit. If the general PSK or QAM modulation is combined with OFDM, each subcarrier corresponds to more than one bit and another set of signals $\sin(2\pi(f_c + k\Delta f)t)$ should be included in Equation (5-73). Thus the sent signal is expressed as

$$s(t) = \sum_{k=-N/2}^{N/2-1} \left(m_{k+N/2,1} \cos(2\pi(f_c + k\Delta f)t) + m_{k+N/2,2} \sin(2\pi(f_c + k\Delta f)t) \right) \tag{5-74}$$

for $0 < t < T$. Note that the form in the main bracket of Equation (5-74) is actually similar to the form of the QPSK signal expressed in Equation (5-67) or QAM signal expressed in Equation (5-69). Therefore, we may view each subcarrier as a QPSK or QAM signal. That is

$$m_{k+N/2,1} \cos(2\pi(f_c + k\Delta f)t) + m_{k+N/2,2} \sin(2\pi(f_c + k\Delta f)t)$$

now corresponds to more than one bit. For example, the signal vector $(m_{k+N/2,1}, m_{k+N/2,2})$ may be from the signal constellations QPSK as shown in Figure 5-21, 8PSK as shown in Figure 5-22, or 16QAM as shown in Figure 5-24. For simplicity, we will consider only BPSK modulation in the following.

What is the advantage of mixing, say QPSK with OFDM? Note that if QPSK is used, one frequency can be used to represent two bits. Thus, we may use N frequencies to represent 2N bits. If higher order PSK or QAM is used, N frequencies can represent even more bits.

Equation (5-73) is very similar to Equation (3-175). For the discrete Fourier transform, we are given $x(0)$, $x(1)$, $\ldots$, $x(n - 1)$ and our task is to find $A_0, A_1, \cdots, A_{n-1}$. In our case, the task is reversed. In one sense, we are given $A_0, A_1, \cdots, A_{n-1}$ and our aim is to find $s(t)$. If we compare Equations (5-73) and (3-175), we notice that the finding of $s(t)$ is similar to performing an inverse discrete Fourier transform. Actually, this is true and will be shown later in this section. *That is, by performing an inverse discrete Fourier transform, we can obtain s(t). This is why the OFDM system can be efficiently implemented.*

Let $\phi_k(t) = \sqrt{2/T}\cos(2\pi(f_c + k\Delta f)t)$, for $0 < t < T$ and $k = -N/2, \ldots, (N/2) - 1$. We can show that $\Delta f = 1/T$ is the smallest channel spacing so that $\phi_i(t)$ and $\phi_j(t)$, for $i \neq j$, are orthogonal. To verify the orthogonality between $\phi_i(t)$ and $\phi_j(t)$, we have

$$
\begin{aligned}
&\langle \phi_i(t), \phi_j(t) \rangle \\
&= \frac{2}{T} \int_o^T \cos(2\pi(f_c + i\Delta f)t) \cos(2\pi(f_c + j\Delta f)t)\,dt \\
&= \frac{2}{T} \int_o^T \frac{1}{2}(\cos(2\pi(i-j)\Delta ft) + \cos(2\pi(2f_c + (i+j)\Delta f)t))\,dt \\
&= \begin{cases} 1 & \text{for } i = j \\ 0 & \text{for } i \neq j. \end{cases}
\end{aligned}
\tag{5-75}
$$

In the above proof, $f_c = c/T$ and $\Delta f = 1/T$ are both used.

The orthogonality of $\phi_i(t)$ and $\phi_j(t)$ can also be established by observing $f_c = c/T$ and $\Delta f = 1/T$. Let $f_0 = 1/T$. We thus have $\phi_k(t) = \cos(2\pi(c + k)f_0 t)$. By using Equation (2-13), we obtain $\langle \phi_i(t), \phi_j(t) \rangle = 0$ for $i \neq j$.

We have proved that $\phi_i(t)$ and $\phi_j(t)$ are orthonormal for $i \neq j$. Now, for simplicity, assuming that the received signal is received without distortion, we have the received signal

$$
r(t) = s(t) = \sqrt{\frac{T}{2}} \sum_{k=-N/2}^{N/2-1} m_{k+N/2} \phi_k(t) \quad \text{for} \quad 0 < t < T.
\tag{5-76}
$$

The problem now is how to demodulate the binary message $m_{j+N/2}$ carried by the jth subcarrier. The solution is to pass the received signal through a correlator that correlates the signal $\phi_j(t)$. The output of the jth correlator can then be expressed as

$$
\begin{aligned}
x_{j+N/2} &= \int_0^T r(t)\phi_j(t)\,dt \\
&= \int_0^T \sqrt{\frac{T}{2}} \sum_{k=-N/2}^{N/2-1} m_{k+N/2}\phi_k(t)\phi_j(t)\,dt \\
&= \sqrt{\frac{T}{2}} \sum_{k=-N/2}^{N/2-1} m_{k+N/2} \int_0^T \phi_k(t)\phi_j(t)\,dt \\
&= \sqrt{\frac{T}{2}} m_{j+N/2}.
\end{aligned}
\tag{5-77}
$$

The last equality follows since $\phi_k(t)$ and $\phi_j(t)$ are orthonormal for $k \neq j$. Therefore, all information bits contained in all subcarriers can be separated at the receiver and hence can be detected independently. Figure 5-25 illustrates a simple OFDM system with four carriers. The carrier frequency f_c is set to be 3 Hz and Δf is set to be 1 Hz. The information bits (m_0, m_1, m_2, m_3) are BPSK-modulated to the carriers with 1, 2, 3 and 4 Hz, respectively. The resulting signals from each subcarrier are combined and

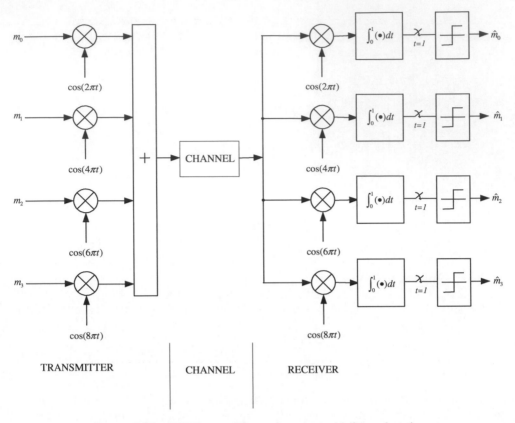

Figure 5-25   OFDM transmitter and receiver with four subcarriers

transmitted over the channel. At the receiver, a bank of correlators are employed to separate the information bit carried by each subcarrier. The output signal from each correlator is compared with a threshold. The threshold is set to be zero in this case. If the output signal is greater than zero, a decision of 1 is made; otherwise a decision of 0 is made.

Figure 5-26 shows all of the four carriers for $0 \leq t < 1$. In this example, all subcarriers have the same phase and amplitude, but in practice the amplitude and phase may be modulated differently for each subcarrier. Figure 5-27 shows all the possible waveforms as (m_0, m_1, m_2, m_3) ranging over all the 16 possible combinations of values.

If the OFDM system is implemented as shown in Figure 5-25, N oscillators have to be used. This may be quite difficult in practice. Actually, the OFDM modulator and the demodulator can be easily implemented by discrete Fourier transform (DFT), and hence the fast Fourier transform (FFT) introduced in Chapter 3 can be employed in OFDM systems. Essentially, we can point out that one of the steps in modulating the input signal can be implemented by using the inverse discrete Fourier transform technique.

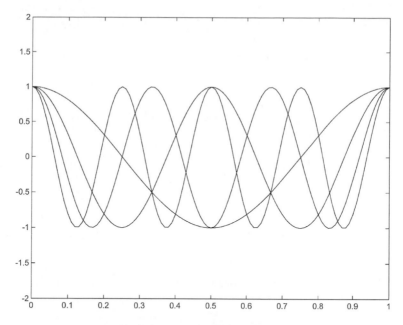

Figure 5-26   Some OFDM carrier signals

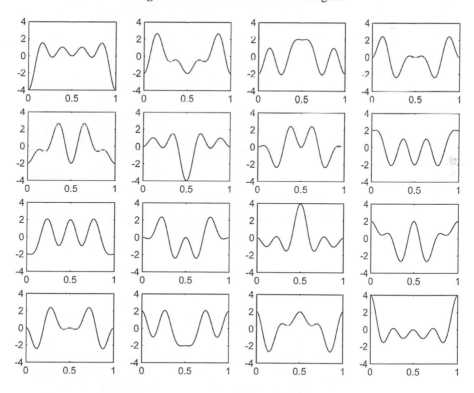

Figure 5-27   All 16 possible combinations of OFDM signals

From Equation (5-73), recalling that $\Delta f = 1/T$, we have

$$
\begin{aligned}
s(t) &= \sum_{k=-N/2}^{N/2-1} m_{k+N/2} \cos(2\pi(f_c + k\Delta f)t) \\
&= \sum_{k=-N/2}^{N/2-1} m_{k+N/2} \cos\left(2\pi\frac{k}{T}t + 2\pi f_c t\right) \quad \text{for} \quad 0 < t < T. \\
&= \operatorname{Re}\left(\sum_{k=-N/2}^{N/2-1} m_{k+N/2} e^{j\frac{2\pi kt}{T}} e^{j2\pi f_c t}\right).
\end{aligned}
\tag{5-78}
$$

For QPSK or QAM modulation, from Equation (5-74), we may have

$$
\begin{aligned}
s(t) &= \sum_{k=-N/2}^{N/2-1} (m_{k+N/2,1} \cos(2\pi(f_c + k\Delta f)t) + m_{k+N/2,2} \sin(2\pi(f_c + k\Delta f)t)) \\
&= \sum_{k=-N/2}^{N/2-1} \left(m_{k+N/2,1} \cos\left(2\pi\frac{k}{T}t + 2\pi f_c t\right) + m_{k+N/2,2} \sin\left(2\pi\frac{k}{T}t + 2\pi f_c t\right)\right) \\
&= \operatorname{Re}\left(\sum_{k=-N/2}^{N/2-1} m_{k+N/2} e^{j\frac{2\pi kt}{T}} e^{j2\pi f_c t}\right),
\end{aligned}
\tag{5-79}
$$

where $m_{k+N/2} = m_{k+N/2,1} - jm_{k+N/2,2}$ is a complex number. All of the points $m_{k+N/2}$ can be plotted on a complex plan which will be similar to the signal constellation employed. Therefore, for QPSK or QAM modulation, the message $m_{k+N/2}$ is a complex number. Let

$$
s'(t) = \sum_{k=-N/2}^{N/2-1} m_{k+N/2} e^{j\frac{2\pi kt}{T}}.
\tag{5-80}
$$

The term $s'(t)$ is called the **equivalent baseband signal** of $s(t)$ because no carrier frequency f_c is involved. Then

$$
s(t) = \operatorname{Re}(s'(t)e^{j2\pi f_c t}).
\tag{5-81}
$$

Let

$$
s'(t) = s'_1(t) + js'_2(t).
\tag{5-82}
$$

It should be noted that $s'(t)$ is a complex function of time which is artificially generated. We can now see that

$$
\begin{aligned}
s(t) &= \mathrm{Re}(s'(t)e^{j2\pi f_c t}) \\
&= \mathrm{Re}[(s'_1(t)+js'_2(t))(\cos(2\pi f_c t)+j\sin(2\pi f_c t))] \\
&= \mathrm{Re}[(s'_1(t)\cos(2\pi f_c t)-s'_2(t)\sin(2\pi f_c t))+j(s'_2(t)\cos(2\pi f_c t)+s'_1(t)\sin(2\pi f_c t))] \\
&= s'_1(t)\cos(2\pi f_c t)-s'_2(t)\sin(2\pi f_c t).
\end{aligned}
\tag{5-83}
$$

Equation (5-83) shows that $s(t)$ can be obtained after $s'(t)$ is obtained. The term $s'(t)$ can be obtained by a straightforward implementation. But this requires the multiplication of cosine and sines by hardware, which is by no means easy. In the following, we shall introduce a method to obtain $s'(t)$ efficiently.

Note that $s'(t)$ is defined in Equation (5-80). In that equation the m_i have already been given and our job is to compute $s'(t)$ for all t. A look at Equation (5-80) will remind us that we are dealing with a case which is quite similar to the inverse discrete Fourier transform introduced in Sections 3.8 and 3.9. Thus, our method to find $s'(t)$ will consist of the following two stages:

1. Instead of computing $s'(t)$ for all t, we only compute $s'(t)$ for N points. That is, for $n = 0, 1, 2, \cdots, N-1$, we compute the following points:

$$
s'_i = s'\left(\frac{iT}{N}\right) \quad \text{for } i = 0, 1, 2, \cdots, N-1.
\tag{5-84}
$$

Later, we shall prove that the computation of the s'_i is equivalent to computing an inverse discrete Fourier transform.
2. After we have obtained $s'_0, s'_1, \ldots, s'_{N-1}$, we can perform an interpolation to obtain $s'(t)$ for all t.

Let us go back to the problem of computing $s'_0, s'_1, \ldots, s'_{N-1}$. From Equations (5-84) and (5-80) we have

$$
\begin{aligned}
s'_i &= s'\left(\frac{iT}{N}\right) \\
&= \sum_{k=-N/2}^{N/2-1} m_{k+N/2} e^{j2\pi k\left(\frac{iT}{N}\right)\frac{1}{T}} \\
&= \sum_{k=-N/2}^{N/2-1} m_{k+N/2} e^{j\frac{2\pi ki}{N}} \\
&= \sum_{k=0}^{N/2-1} m_{k+N/2} e^{j\frac{2\pi ki}{N}} + \sum_{k=-N/2}^{-1} m_{k+N/2} e^{j\frac{2\pi ki}{N}}.
\end{aligned}
\tag{5-85}
$$

Let $k' = k + N$; the second term of the Equation (5-85) can be expressed as

$$\sum_{k=-N/2}^{-1} m_{k+N/2} e^{j\frac{2\pi ki}{N}}$$

$$= \sum_{k'=N/2}^{N-1} m_{k'-N/2} e^{j\frac{2\pi (k'-N)i}{N}}$$

$$= \sum_{k'=N/2}^{N-1} m_{k'-N/2} e^{j\frac{2\pi k' i}{N}} e^{j\frac{2\pi (-N)i}{N}} \qquad (5\text{-}86)$$

$$= \sum_{k'=N/2}^{N-1} m_{k'-N/2} e^{j\frac{2\pi k' i}{N}} e^{-j2\pi i}$$

$$= \sum_{k'=N/2}^{N-1} m_{k'-N/2} e^{j\frac{2\pi k' i}{N}}.$$

The last equality follows since $e^{-j2\pi i} = 1$. Now, we may rewrite (5-85) as

$$s_i' = \sum_{k=0}^{N/2-1} m_{k+N/2} e^{j\frac{2\pi ki}{N}} + \sum_{k=N/2}^{N-1} m_{k-N/2} e^{j\frac{2\pi ki}{N}}. \qquad (5\text{-}87)$$

We define

$$A_k' = \begin{cases} m_{k+N/2} & \text{for} \quad 0 \leq k \leq N/2 - 1 \\ m_{k-N/2} & \text{for} \quad N/2 \leq k \leq N - 1. \end{cases} \qquad (5\text{-}88)$$

We have

$$s_i' = \sum_{k=0}^{N-1} A_k' e^{j\frac{2\pi ki}{N}}. \qquad (5\text{-}89)$$

From Equation (5-89) we can see that finding the s_i' is equivalent to computing the inverse discrete Fourier transform of A_i' as discussed in Section 3.9, where A_i' is defined in Equation (5-88) and is known because the m_i are known to us. Therefore, the discrete-time equivalent of the modulated signal $s'(t)$ can be obtained by performing an inverse discrete Fourier transform of A_i'. In practice, this transform can be implemented very efficiently by the inverse fast Fourier transform (IFFT). The IFFT drastically reduces the number of calculations by exploiting the regularity of the exponential functions.

Let us summarize the above discussion as follows:

1. Our major job is to find $s(t)$ as defined in Equation (5-78):

$$s(t) = \text{Re} \left(\sum_{k=-N/2}^{N/2-1} m_{k+N/2} e^{j\frac{2\pi kt}{T}} e^{j2\pi f_c t} \right).$$

2. We define $s'(t)$ according to Equation (5-80) as follows:

$$s'(t) = \sum_{k=-N/2}^{N/2-1} m_{k+N/2} e^{j\frac{2\pi kt}{T}}$$

and let

$$s'(t) = s_1(t) + js_2(t)$$

as in Equation (5-82). Then, according to Equation (5-83):

$$s(t) - s_1'(t)\cos(2\pi f_c t) \quad s_2'(t)\sin(2\pi f_c t).$$

3. To find $s'(t)$, we first obtain $A_0', A_1', \ldots, A_{N-1}'$ from $m_0, m_1, \cdots, m_{N-1}$ according to Equation (5-88) as follows:

$$A_k' - \begin{cases} m_{k+N/2} & \text{for} \quad 0 < k \le N/2 - 1 \\ m_{k-N/2} & \text{for} \quad N/2 \le k \le N - 1. \end{cases}$$

4. We compute the inverse discrete Fourier transform of $A_0', A_1', \ldots, A_{N-1}'$ and obtain $s_0', s_1', \ldots, s_{N-1}'$, where $s_i' = s'(iT/N)$, by using Equation (5-89).
5. Having obtained $s_0', s_1', \ldots, s_{N-1}'$, we perform an interpolation on them to obtain $s'(t)$.
6. Having obtained $s'(t)$, we can use Equation (5-83) to obtain $s(t)$. That is, the following equation is used:

$$s(t) = s_1'(t)\cos(2\pi f_c t) - s_2'(t)\sin(2\pi f_c t)$$

Equation (5-83) can be obtained from another perspective. Let us start with Equation (5-73):

$$s(t) = \sum_{k=-N/2}^{N/2-1} m_{k+N/2} \cos(2\pi(f_c + k\Delta f)t) \quad \text{for} \quad 0 < t < T.$$

Thus:

$$s(t) = \sum_{k=-N/2}^{N/2-1} m_{k+N/2} \cos(2\pi(f_c + k\Delta f)t)$$

$$= \sum_{k=-N/2}^{N/2-1} m_{k+N/2}(\cos(2\pi f_c t)\cos(2\pi k\Delta ft) - \sin(2\pi f_c t)\sin(2\pi k\Delta ft))$$

$$= \cos(2\pi f_c t) \sum_{k=-N/2}^{N/2-1} m_{k+N/2}\cos(2\pi k\Delta ft) - \sin(2\pi f_c t) \sum_{k=-N/2}^{N/2-1} m_{k+N/2}\sin(2\pi k\Delta ft).$$

$$(5\text{-}90)$$

Now let

$$s'(t) = \sum_{k=-N/2}^{N/2-1} m_{k+N/2}e^{j\frac{2\pi kt}{T}}.$$

We can easily see that

$$s'(t) = \sum_{k=-N/2}^{N/2-1} m_{k+N/2}\cos(2\pi k\Delta ft) + j \sum_{k=-N/2}^{N/2-1} m_{k+N/2}\sin(2\pi k\Delta ft). \qquad (5\text{-}91)$$

If we let $s'(t) = s'_1(t) + js'_2(t)$ we have:

$$s'_1(t) = \sum_{k=-N/2}^{N/2-1} m_{k+N/2}\cos(2\pi k\Delta ft)$$

$$(5\text{-}92)$$

$$s'_2(t) = \sum_{k=-N/2}^{N/2-1} m_{k+N/2}\sin(2\pi k\Delta ft).$$

Substituting Equation (5-92) into (5-90), we obtain Equation (5-83).

Note that Equation (5-78) involves a large number of multiplications and it is complicated to implement it by hardware. The fast inverse discrete Fourier transform used in the above procedure, on the other hand, is much easier to implement, as discussed in Chapter 3.

The modulator using the techniques discussed above is shown in Figure 5-28(a). Interpolation is used. Since the output of the IFFT is a discrete-time signal, it has to be converted to a continuous-time signal by a parallel-to-serial converter followed by a digital-to-analog (D/A) converter.

In the receiver, the received signal is $s(t)$. To demodulate, we have to find $s'_1(t)$ and $s'_2(t)$. To do this, we note that, according to Equation (5-83):

$$s(t) = s'_1(t)\cos(2\pi f_c t) - s'_2(t)\sin(2\pi f_c t).$$

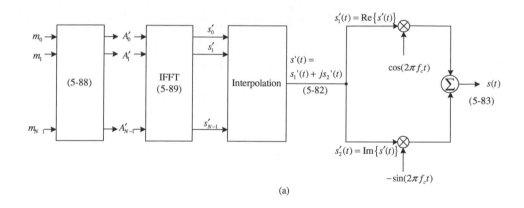

(a)

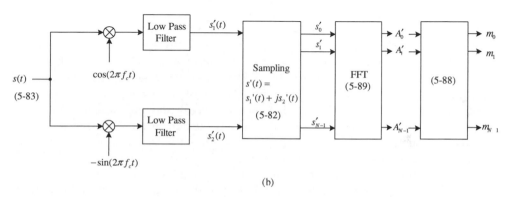

(b)

Figure 5-28 Simple OFDM system with IFFT and FFT: (a) modulator, and (b) demodulator

To find $s'_1(t)$, we multiply $s(t)$ by $\cos(2\pi f_c t)$. The result is

$$
\begin{aligned}
s(t)\cos(2\pi f_c t) &= s'_1(t)\cos^2(2\pi f_c t) - s'_2(t)\sin(2\pi f_c t)\cos(2\pi f_c t) \\
&= s'_1(t)\frac{1}{2}(1 + \cos(4\pi f_c t)) - s'_2(t)\frac{1}{2}\sin(4\pi f_c t).
\end{aligned}
\tag{5-93}
$$

In Equation (5-93), $(1/2)s'_1(t)$, which is of low frequency, appears. There are other terms which are all of high frequency. Thus $s'_1(t)$ can be obtained by using a low-pass filter. This is illustrated in Figure 5-28(b). $s'_2(t)$ can be obtained similarly by multiplying $s(t)$ by $\sin(2\pi f_c t)$. After $s'_1(t)$ and $s'_2(t)$ are found, we use Equation (5-82) to find $s'(t) = s'_1(t) + js'_2(t)$. We sample $s'_1(t)$ N times to obtain $s'_0, s'_1, \ldots, s'_{N-1}$. We then use FFT to find $A'_0, A'_1, \ldots, A'_{N-1}$ by using Equation (5-89). After getting $A'_0, A'_1, \ldots, A'_{N-1}$, $m'_0, m'_1, \ldots, m'_{N-1}$ can be recovered by using Equation (5-88) and the demodulation is completed.

A naive reader may be puzzled by a problem when the OFDM system is implemented by using the FFT technique. The OFDM system is based upon the orthogonality of the signals. The demodulator uses an inner product to recover the m'_i as shown in Figure 5-25. If FFT is used to implement the OFDM system, as shown in Figure 5-28(b), no inner product is used. What happened? Actually, the inner product is still used. If the traditional method is used, the inner product involves the multiplication of $s(t)$ and $\cos(2\pi k f_c t)$ to find m_k.

Now, FFT is used in the demodulation process. FFT uses the inner product of two vectors

$$\left(s'_0,\ s'_1,\ldots,\ s'_{N-1}\right) \text{ and } \left(e^{-j\frac{2\pi k(0)}{N}},\ e^{-j\frac{2\pi k(1)}{N}},\ldots,\ e^{-j\frac{2\pi k(N-1)}{N}}\right)$$

to find A'_k. Thus, the inner product scheme is still used.

As for the carrier frequency and bandwidth of the OFDM system, note that this system has a set of N carrier frequencies. Each consecutive pair of carrier frequencies is separated by Δf. Thus the total bandwidth is $N\Delta f = N/T$, where T is the symbol duration and is still determined by the data rate as we discussed before. For the OFDM system, $T = NT_b$, where T_b is the bit duration because N bits are sent together. Thus, the bandwidth is $N/T = N/NT_b = 1/T_b$, which is not significantly reduced. But there is one advantage of using the OFDM system. If we do not use the OFDM system and use only one carrier frequency as in M-ary PSK and M-QAM systems, we have to design a rather wide bandwidth system if the bit rate is high. To support the same data rate, the single carrier system must occupy about the same signal bandwidth as the OFDM system. For a practical communication channel, the frequency response of the channel is not flat. If a single carrier frequency is used, the receiver has to deal with a wideband signal, of which the signal is distorted due to the non-flatness of the channel response. For each frequency in the band, if the signal associated with it is attenuated, for a proper demodulation, the receiver has to compensate it. The compensation of the signal is called 'equalization' and the device performing such a task is called an 'equalizer'. For the signal carrier system, the equalizer is generally implemented using a specially designed filter. This is considered to be difficult in general. On the other hand, for the OFDM system, there are N subchannels. For each subchannel, there is only a frequency with a very narrow band. Therefore, each subchannel can be considered as having the same attenuation. So far as the receiver is concerned, within the entire bandwidth, it only has to make sure that it can recover N signals with distinct frequencies and very narrow bandwidths, which is much easier.

There is another advantage of using the OFDM system. Note that in this system, N bits are sent together. Thus its symbol duration is quite long as N is usually quite large. If only one bit is sent at a time, the symbol duration is much shorter. It is easier to cause intersymbol interference for short symbol durations than for long symbol durations because we can put some guard signals at the beginning of a symbol for long symbol durations. For short symbol durations, it is harder to put a guard signal at the beginning of a symbol because its duration is very short.

Now, the reader may wonder whether using the M-ary PSK system or the M-QAM system with single carrier may achieve such a long symbol duration. For OFDM, N can be as large as 256, as in the ADSL system. But it is absolutely impossible for the former system to have 256 bits sent together because 2^{256} is too large to be practical.

ADSL: An Application of the OFDM Technology

ADSL (Asymmetrical Digital Subscriber Line) is widely used these days to connect computers to the Internet. This system uses the Discrete Multitone Technology (DMT) which is quite similar to OFDM. There are 256 subchannels. Subchannel spacing Δf is 4.312 kHz. The lowest frequency is also 4.312 kHz. The total bandwidth is 1.104 MHz. As suggested by the name, ADSL provides uploading and downloading which are different. The uploading data rate ranges from 160 to 784 KBit/s (bps) and the downloading data rate ranges from 1.544 to 8 Mbps. There are two types for the number of subchannels being used:

- Type 1: The uploading uses 1 to 32 subchannels and the downloading uses 33 to 256 subchannels.
- Type 2: The uploading uses 1 to 32 subchannels and the downloading uses 1 to 256 subchannels.

The multi-user problem of ADSL is handled at a higher level. That is, data are routed to different users by using the IPs of these computers.

5.8 OFDM in Wireless Local Area Networks

Orthogonal frequency division multiplexing (OFDM) has been found to have many advantages, such as the robustness to the intersymbol interference (ISI) and multipath fading (which will not be discussed in this book). Therefore, OFDM has been adopted in IEEE 802.11 as a standard for wireless local area networks (WLANs). In this section, we will briefly describe the applications of OFDM in the WLAN standard.

The IEEE 802.11a supports data rates of 6, 9, 12, 18, 24, 36, 48 and 54 Mbps. Different data rates are achieved by using different convolutional codes and different types of modulation. In this section, we will focus on the types of modulation. The convolutional codes will be introduced in Chapter 8. The carrier frequency is 5 GHz.

For IEEE 802.11a, the total number of subcarriers is $N = 64$. Among the 64 subcarriers, only 52 are activated and the other subcarriers are set to be zero. From Equation (5-80), the indices of the activated subcarriers are from -26 to $+26$, excluding the index 0. Therefore, from Equation (5-88), only $A'_1, A'_2, \ldots, A'_{26}, A'_{38}, A'_{39}, \ldots, A'_{63}$ carry information and the others are set to be zero. The channel spacing $\Delta f = 1/T = 312.5$ kHz. Therefore, the total bandwidth of the OFDM signal is about 312.5 kHz $\times 26 \times 2 = 15.250$ MHz. The A'_i are complex numbers for QPSK, 16-QAM or 64-QAM systems as pointed out in Section 5.7 immediately after the presentation of Equation (5-79) and depend on the data rate. The signal constellations for these systems are shown in Figure 5-29. The real part of A'_i is represented by the horizontal axis and the imaginary part is represented by the vertical axis. For example, in 16-QAM, for the signal point indexed by 1010, the value of A'_i should be $3 + 3j$. The discrete-time equivalent of the modulated signal $s'(t)$ can be obtained by performing an inverse discrete-time Fourier transform of A'_i as shown in Equation (5-89). This transform can be implemented efficiently by the inverse fast Fourier transform (IFFT). The outputs of the IFFT are discrete time signals, denoted as $s_0, s_2, \ldots, s_{63}$, which are complex numbers. The real parts of $s_0, s_2, \ldots, s_{63}$ are used to modulate the cosine waveform and the imaginary part is used to modulate the sine waveform as shown in Equation (5-83).

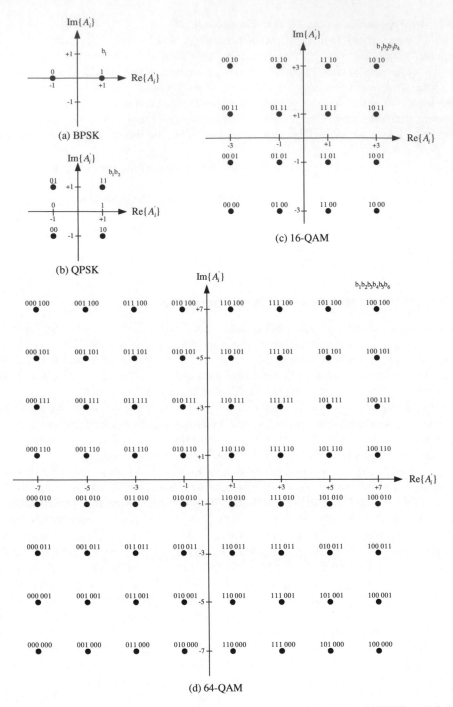

Figure 5-29   Signal constellations for IEEE 802.11a standard: (a) BPSK; (b)QPSK; (c)16-QAM; (d) 64-QAM

The discrete signals are then converted to continuous-time by a parallel-to-serial converter followed by a digital-to-analog (D/A) converter with sampling rate N/T. Finally, the signal is up-converted to the carrier frequency. For IEEE 802.11a, the carrier frequency of the system is around 5 GHz. The modulated signal is then amplified by a power amplifier (PA) and transmitted via the antenna. The up-conversion is accomplished by analog modulation introduced in Chapter 4.

At the receiver, since the signal may suffer serious attenuation, a low-noise amplifier (LNA) is employed to amplify the received signal. The signal is then down-converted to the baseband equivalent signal by multiplying the received signal with cosine and sine waveforms with frequency f_c, as discussed in Section 5.7. The resulting signal is $s'(t)$ for the distortion-less case. Notice that $s'(t)$ is also a complex number. The real part of $s'(t)$ is the signal from the upper branch and the imaginary part is the signal from the lower branch as shown in Figure 5-28. The resulting signal is then sampled by analog-to-digital (A/D) converters with sampling rate N/T. A serial-to-parallel converter is then employed to store the samples, resulting in the samples $\left\{s'_i\right\}_{i=0}^{63}$. Finally, an FFT is employed to recover $\left\{A'_k\right\}_{k=0}^{63}$ from $\left\{s'_i\right\}_{i=0}^{63}$ by taking the FFT of $\left\{s'_i\right\}_{i=0}^{63}$.

The IEEE 802.11a supports multiple users. The mechanism to handle the multiple access problem is to use the carrier-sense multiple access (CSMA) technique which will be introduced in the next chapter. There is also an IEEE 802.11 g system which is quite similar except that its carrier frequency is 2.4 GHz.

5.9 Digital Audio Broadcast Using OFDM and TDMA

The digital audio broadcast (DAB) system is a new multimedia broadcast system which will replace the existing AM and FM audio broadcast services in the future. It was developed in the 1990s by the Eureka 147/DAB project. DAB provides high-quality digital audio that is comparable to that of CD quality and is able to transmit program-associated data (e.g. song titles and artists) and a multiplexing of other data services (e.g. traffic and travel information, still and moving pictures). DAB uses what are called *multiplexes* to group radio stations together. The technology is beyond this book and will not be discussed. In this way, several audio programs and data services can be multiplexed into the same frequency band and transmitted simultaneously.

A simplified DAB transmitter is shown in Figure 5-30. The DAB system combines two major technologies to provide high-quality audio service and robust reception of the wireless signal. The first technology is audio compression and the second technology is OFDM. The audio compression performed by the block 'audio encoder' as shown in Figure 5-30 can efficiently reduce the required bit rate while maintaining good audio quality. The OFDM technology was introduced earlier and has been found to have many advantages, such as robustness to intersymbol interference (ISI). Therefore, the OFDM system can provide robust reception of wireless signals. Another very important technology that DAB uses is the TDMA mechanism which will be introduced in the next chapter.

Three major data services are transmitted by the DAB transmitter – audio, data and service information. The audio and data are encoded and transmitted through a main service channel (MSC) multiplexer. The MSC uses the TDMA technology which will be introduced in Chapter 6. The interleaved data channel is divided into a number of subchannels

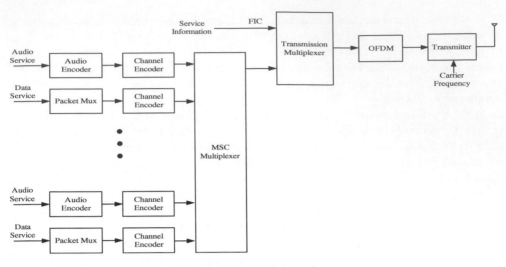

Figure 5-30 DAB transmitter

which are individually encoded by convolutional encoder, which will be introduced in Chapter 8 – denoted 'channel encoder' in Figure 5-30. The service information is transmitted through a fast information channel (FIC) which provides rapid access to information by the receiver. For example, the receiver must know how different services are multiplexed in advance to properly de-multiplex, and locates the services that it wants to receive. The service information will provide such multiplex configuration information (MCI) to the receiver.

DAB provides several transmission modes: TM I, TM II, TM III and TM IV for different environments. Table 5-5 gives the major parameters of the OFDM system.

We will not go into the detail of the DAB standard and focus only on the modulation techniques. Denote the total number of subcarriers in the OFDM system as N. Among the N subcarriers, only K are activated and the others are set to be zero. As given in Table 5-5, N and K are different for different transmission modes. In addition, to reduce the cost of the DAB receiver, the DAB system uses a very simple modulation technique, called Differential Quadriphase-shift Keying (DQPSK) which is a variation of QPSK (see Section 5.5). The difference between DQPSK and QPSK is that, for DQPSK, the two information bits (m_1, m_2) are encoded into the *difference* of the phases between previous transmitted symbol and current transmitted symbol; while for QPSK the two information bits (m_1, m_2) are directly mapped to the phase of the current transmitted symbol and are independent of the previous

Table 5-5 Transmission modes of DAB and the corresponding OFDM parameters

Mode	N	K	$\Delta f = 1/T$
TM I	2048	1536	1 kHz
TM II	512	384	4 kHz
TM III	256	192	8 kHz
TM IV	1024	768	2 kHz

symbol. To recover the original message (m_1, m_2), the receiver just needs to compare the phase difference between two consecutive received symbols.

In Table 5-5, T is the symbol duration and $\Delta f = 1/T$ is the bandwidth of each subchannel. Therefore the total bandwidth of the OFDM signal is about $k\Delta f = 1.5\,\text{MHz}$ for all transmission modes.

Since the DAB system uses OFDM technology, there are several frequency bands employed by the DAB standard. For example:

- Band 1: 47.936 MHz to 67.072 Mhz, channel spacing about 1.7 MHz
- Band II: 174.928 MHz to 239.200 MHz, channel spacing about 1.7 MHz
- L-Band: 154.960 MHz to 1491.184 MHz, channel spacing about 1.7 MHz.

Since the digital audio broadcasting system employs TDMA technology, a digital radio station may use one frequency to broadcast several programs. That is, a user may tune to one station and choose one of the programs to listen to. This kind of service is thus better than the traditional one where one station can provide only one program. That DAB can do this is due to the fact that the data are digital and it is easier to implement the TDMA technology when the data are in digital form.

5.10 The Role of Inner Product in Digital Modulation

In the above sections, although different modulation techniques have been introduced, there is actually a common basic principle: to use the inner product concept. Consider the binary frequency shift keying (FSK) system, for instance. In this case, we have two signals: $\cos 2\pi f_1 t$ and $\cos 2\pi f_2 t$. At any moment, either we receive $\cos 2\pi f_1 t$ or $\cos 2\pi f_2 t$. Let the received signal be denoted as $y(t)$. To detect which signal was transmitted, we may use the inner product concept. Suppose we multiply $y(t)$ by $\cos 2\pi f_1 t$ and $\cos 2\pi f_2 t$, respectively, and integrate. The result will let us know whether the transmitted signal is 1 or 0. This is the basis of our demodulation mechanism. That our mechanism works is due to the fact that the two signals, namely $\cos 2\pi f_1 t$ and $\cos 2\pi f_2 t$, are othogonal.

Let us now consider the quadriphase-shift keying system (QPSK). In this case, we have a mixed signal as follows:

$$s_i(t) = s_{i1}\sqrt{\frac{2}{T}}\cos(2\pi f_c t) + s_{i2}\sqrt{\frac{2}{T}}\sin(2\pi f_c t).$$

Our task is to detect s_{i1} and s_{i2}. This can be easily done because $\cos(2\pi f_c t)$ and $\sin(2\pi f_c t)$ are orthogonal. To detect s_{i1} we multiply $s_i(t)$ by $\cos(2\pi f_c t)$ and integrate; and to detect s_{i2} we multiply $s_i(t)$ by $\sin(2\pi f_c t)$ and integrate. That this will work is due to the fact that $\cos(2\pi f_c t)$ and $\sin(2\pi f_c t)$ are orthogonal as pointed out in Chapter 3.

Finally, the reader can now easily see why the orthogonal frequency division multiplexing scheme (OFDM) works. Although we have mixed a set of signals with different frequencies, we have made sure that they are orthogonal to one another. To detect the signal with a certain frequency, say $f_c + i\Delta f$, we only have to multiply the received signal by $\cos(2\pi(f_c + i\Delta f)t)$ and integrate. Only the associated coefficient, namely $m_{i+N/2}$, survives and all of the other coefficients vanish.

5.11 Review of Digital Modulation Techniques

Digital modulation is much more complicated than analog modulation. A review of these techniques may be helpful here.

1. Baseband pulse transmission, strictly speaking, has no modulation involved. It cannot be used in a wireless environment.
2. For all of the other digital modulation techniques, digital signals are transformed into analog signals. Each pulse is now transmitted as a truncated sinusoidal signal.
3. For the ASK system, the amplitude of the sinusoidal function is used to indicate whether the bit is 1 or 0.
4. For the PSK system including BPSK, QPSK and M-ary PSK, phase is used to distinguish the values of the bit.
5. For the FSK system, frequency is used to distinguish the values of the bit.
6. For QPSK, M-ary PSK and QAM systems, bits are bundled together to be transmitted. All of the bits are represented by sinusoidal functions with the same frequency. The advantage of such systems is that the bandwidth will be reduced if the same bit rate is used and the bit rate can be increased if the bandwidth is kept the same.
7. For the OFDM system, bits are also bundled together to be transmitted. Each bit is now represented by a sinusoidal function. But the frequencies are all different. This system can be used together with QPSK, M-ary PSK and QAM. The greatest advantage of the system is that the discrete Fourier transform technique can be used. One advantage of using the OFDM system is that it consists of many narrow bandwidth subchannels. It is easier to design a receiver to cope with such narrow bandwidth subchannels. Besides, the symbol duration is much longer and this is another advantage.
8. It should be emphasized here that digital signals are transmitted in purely digital forms only in the baseband pulse transmission. As the reader can see, in all of the other cases, digital signals are modulated into analog signals and transmitted with a carrier frequency. In other words, all of the carrier frequencies are used to distinguish distinct TV channels. This is why digital modulation can make the frequency division multiple access (FDMA) system possible.
9. In wireless systems, pure digital baseband signals cannot be transmitted at all. The primitive baseband pulse transmission method cannot be used and some carrier frequencies are needed. The carrier frequencies may not be high enough for wireless transmission. Thus analog modulation techniques are often used to up-convert the carrier frequency to an RF frequency for transmission. At the receiver, analog modulation may be used to down-convert the incoming RF signals.
10. The bandwidth of any digital modulation system is determined by the data rate. Let the duration of a symbol be T. Then the bandwidth associated with every carrier frequency is $2/T$. If there are N distinct consecutive carrier frequencies used, the bandwidth of the digital modulation system will be $2N/T$. If OFDM is used, the bandwidth is N/T.
11. Note that for QPSK, 8PSK, QAM and OFDM systems, bits are mixed and sent. For a fixed system, if we keep the bit duration the same, the bit rate will not be increased, but the bandwidth will be reduced. If we reduce the bit duration and thus increase the bit rate, the bandwidth will remain the same. In general, we are given a bandwidth and we

Table 5-6 Minimum Euclidean distances of different constellations

Constellation	Number of bits per symbol	Minimum Euclidean distance	Noncoherent demodulation	Figures of constellations
ASK	1	$\sqrt{2E}$	Yes	5–7
BPSK	1	$2\sqrt{E}$	No	5–10
BFSK	1	$\sqrt{2E}$	Yes	5–16
QPSK	2	$\sqrt{2E}$	No	5–21
8PSK	3	$\sqrt{2E\left(1 - \cos\dfrac{2\pi}{8}\right)}$ $= \sqrt{0.586E}$	No	5–22
16-QAM	4	$\sqrt{0.4E}$	No	5–24

are allowed to increase the bit rate by using QPSK, 8PSK and QAM. But there is a limit as to what we can do, as we shall explain below.

12. Now, if we want to transmit more than one bit in a symbol, we require signal constellation of large size. Table 5-6 gives the information about ASK, BPSK, FSK, QPSK, 8PSK and 16-QAM. The number of transmitted bits per symbol are 2, 3 and 4, respectively, for QPSK, 8PSK and 16-QAM. As shown in the table, the minimum Euclidean distance decreases as the size of the constellation increases. This happens because the constellation will become denser as the size of the constellation increases. This means that if we want to transmit more information bits per symbol, the symbol error rate will increase.

When should we use such a constellation of larger size? We will give an example to explain this point. In most two-way communication systems, the receiver will estimate the channel condition. If the channel is very noisy, the receiver will negotiate with the transmitter to reduce the size of constellation so that the symbol error rate will still be acceptable. If the channel is not noisy, then the receiver will negotiate with the transmitter to increase the size of constellation and hence the data rate can be increased for a less noisy channel.

Further Reading

- For more details about baseband pulse transmission, see [LSW68], [BBC87] and [GHW92].
- For a tutorial paper regarding to different modulation techniques, including ASK, FSK and PSK, see [AD62].
- A detailed analysis on the spectra of FSK system can be found in [AS65].
- For a detailed discussion of DQPSK and its optimum receiver design, see [SD92].
- For discussion of OFDM and its applications, see [ZW95], [CL91], [NP00] and [C85].

Exercises

5.1 For the baseband pulse transmission method, there is no carrier frequency involved. Why do we still say that the transmitted digital signal has a bandwidth? What determines the bandwidth?

5.2   Can the baseband pulse transmission method be used to send signals wirelessly?

5.3   Suppose a pulse is modulated by a carrier signal with a high frequency. Would its bandwidth be enlarged? Explain.

5.4   It was pointed out that we may have a noncoherent detector for ASK. Can we have such a detector for BPSK? Explain.

5.5   Consider the QPSK system. Is it true that the bit rate is doubled since two bits are sent simultaneously? Explain.

5.6   For the QPSK system, suppose we want to send 01. Show that $y_1(t)$ is negative for most of the time and $y_2(t)$ is mostly positive.

5.7   State the advantages of using the M-ary PSK and M-QAM systems.

5.8   Consider Equation (5-73):

$$s(t) = \sum_{k=-N/2}^{N/2-1} m_{k+N/2} \cos(2\pi(f_c + k\Delta f)t).$$

Show that $s(t) = s_1'(t)\cos(2\pi f_c t) - s_2'(t)\sin(2\pi f_c t)$. Give $s_1'(t)$ and $s_2'(t)$.

5.9   Use your own words to explain the essential principles about why OFDM can be implemented by inverse discrete Fourier transform.

5.10 Explain why high data rate systems must also be broadband systems?

5.11 We all know that in digital modulation, pulses are modulated by cosine functions. From Chapter 3, we know that the Fourier transform of a cosine function is the delta function. Why does such a modulated signal still have a bandwidth?

5.12 For the FSK system, if you want to have a higher bit rate, will you have two even wider operating frequencies and a wider bandwidth? Explain.

6

Multiple-access Communications

Among communication systems, the simplest kind consists of one pair of users: one transmits and one receives. This one-to-one communication is traditionally called point-to-point communication.

In many applications, it is required that several transmitters send information simultaneously through the same communication channel. Nowadays, there are numerous examples of multiple access communication in which several transmitters share a common channel. For example, in a mobile telephone system, the ground area is partitioned into several small areas. The partitioned area is called a 'cell', and such a system is called a ***cellular telephony system***. A base station is placed in the center of each cell to provide the communication services of all mobile phone users over the area covered by the radio waves. Now, several mobile phone users may need to communicate with the same base station at the same time. Therefore, some communication techniques are required to solve the multiple access problem.

There are some other examples of multiple access communications: ground stations communicating with a satellite, local-area network, packet-radio network, and cable television network, to name a few. A common feature of those communication channels is that the receiver obtains the superposition (summation) of the signals sent by the active transmitters. This phenomenon causes a less than ideal effect. For example, in wired telephone systems, it causes *crosstalk* between different telephone links. In cellular telephony systems, it causes *serious interference* between users, if the same radio frequency is used simultaneously by distinct users.

Figure 6-1 depicts the multiple access communication scenario: only the case of K users as transmitters and a common receiver is shown. In practice, there may be many receivers, and each is interested in the information sent by one user (or a subset of users) only. For example, each user may contain a transmitter and a receiver that may transmit and receive information via the common channel, or some users play only the role of transmitter and the others play the role of receiver. At one extreme, the same information is delivered to all recipients, for example, in radio and television broadcasting or in cable television. In these broadcast systems, the broadcast stations play the role of transmitter and the TV sets or radio receivers play the role of receiver. At the other extreme, the messages can be transmitted to and received from different users independently. For example, a base station transmits to mobile units and mobile units transmit to a base station. In this case, each user must contain a receiver and a transmitter.

Communications Engineering.   R.C.T. Lee, Mao-Ching Chiu and Jung-Shan Lin
© 2007 John Wiley & Sons (Asia) Pte Ltd

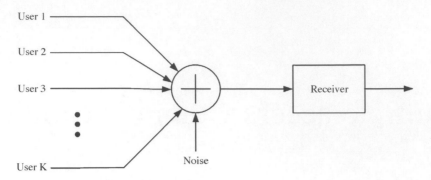

Figure 6-1   Multiple access communication

In this chapter, we first consider the scenario shown in Figure 6-1. The receiver must be capable of recovering the information sent by one user (or a subset of users). Note that the signal received by the receiver is mixed. Therefore, for example, for a base station of a cellular telephony system, the receiver must be capable of recovering all information sent from all the active users; and, for a particular mobile unit, it is sufficient to recover the information that is sent from the base station to that particular user.

Several multiple access techniques will be introduced in this chapter, including the frequency-division multiple access (FDMA), time-division multiple access (TDMA), code-division multiple access(CDMA) and carrier-sense multiple access (CSMA). We assume that the information is already in digital form, and only digital multiple access techniques are considered.

6.1 Frequency-division Multiple Access (FDMA)

Radio-frequency modulation enables several transmissions to coexist in time and space without mutual interference by using different carrier frequencies. For example, for radio or television broadcast systems, several broadcast stations in an area are assigned different radio frequencies so that the signal spectra from the stations do not overlap. A radio or television set can be tuned to receive a particular program by adjusting a band-pass filter inside the set. The band-pass filter passes signals only around a particular center frequency and rejects the others, and the resulting signal can be demodulated without interference from other stations. Currently, information transmitted over most of the radio and television broadcast systems is in analogue form. However, digital broadcast systems that provide better quality of audio and video will become popular in the near future.

An example of a frequency-division multiple-access system is given in Figure 6-2 where the message is assumed to be in digital form. This example may be considered as an uplink scenario for a mobile phone system, where all K users want to transmit messages to the base station. *As shown, in an FDMA system, all of the K active users are assigned with different frequency bands with central frequencies $f_1, f_2, \ldots, f_K$ before transmission.*

Each user then occupies the assigned frequency band during the lifetime of the connection. For the kth user, the message m_k is digitally modulated to the assigned frequency band f_k. Then, by a power amplifier and an antenna, the modulated signal is sent over the air

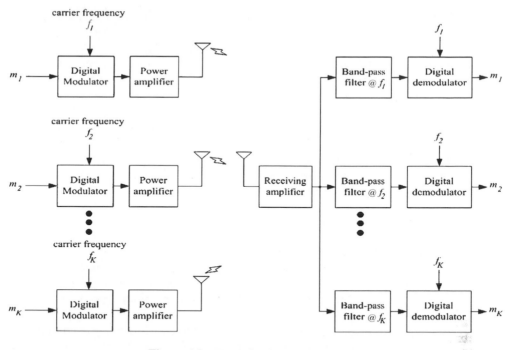

Figure 6-2   Example of an FDMA system

as an electromagnetic (EM) wave. The EM wave will propagate in space to the destination where the receiver resides. All transmitted signals from all the users will appear on the receiving antenna. At the receiver, the signal appearing on the antenna is the superposition of all transmitted signals from all active users. However, since all active users are assigned with different frequency bands, the users' transmitted signals do not overlap in the frequency domain. Figure 6-3 illustrates the spectrum of the received signal on the antenna.

Since the EM waves may be seriously attenuated over long-distance propagation, a receiving amplifier is required to increase the strength of the received signal. The received signal is then fed to K band-pass filters with central frequencies $f_1, f_2, \ldots, f_K$. For example,

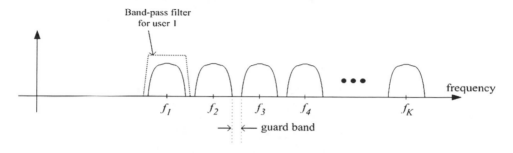

Figure 6-3   Spectrum of an FDMA system

the band-pass filter for the first user can only pass the signal around the central frequency f_1 with proper bandwidth as shown in Figure 6-3, and reject the others. Therefore, the output signal of the first band-pass filter contains only the transmitted waveform of the first user without interference from the others. The digital demodulator then recovers the desired information m_1. Because of the non-ideal effect of the band-pass filter, we may have to insert guard bands in FDMA as shown in Figure 6-3.

All the digital modulation and demodulation techniques introduced in Chapter 5 may be used in this system. However, there are many practical limitations for a mobile phone system, such as a limited number of frequency bands and limited power consumption of the mobile handset. Therefore, evaluation of the various modulation techniques is required.

FDMA technology can also be used for analogue signals. In Chapter 4, we pointed out that both AM and FM broadcasts use FDMA to distinguish users.

6.2 Time-division Multiple Access (TDMA)

In an FDMA system, the frequency domain is partitioned into small non-overlapping bands so that each user's message can be transmitted over the assigned band without interference from other users. *In time-division multiple access (TDMA) systems, all users use the same frequency band, but the time domain is partitioned into slots assigned to each user as shown in Figure 6-4.*

User 1 can transmit data only in the time slots assigned to user 1, user 2 can transmit data only in the time slots assigned to user 2, and so on. Note that FDMA systems allow completely uncoordinated transmissions in the time domain: no time synchronization among the users is required. This advantage is not shared by TDMA systems where all transmitters and receivers must have access to a common clock. An important feature of FDMA and TDMA techniques is that the various users are operating in separate non-interfering channels. Besides, since channels, transmitters and receivers are not ideal, we may require the insertion of guard times between TDMA time slots as shown in Figure 6-4.

An example of a time-division multiple access system is given in Figure 6-5. This may be considered an uplink scenario for a mobile phone system, where all K active users want to transmit messages to the base station. All active users in this system use the same frequency band with central frequency f_c but different time slots according to Figure 6-4. The first user transmits the message in the first slot, the second user transmits the message in the second slot, and so on. By a power amplifier and an antenna, the modulated signal is sent over the air by means of electromagnetic waves. For a particular user, the transmitter can go into low-power mode during the time interval of non-owing slots, thus reducing power consumption at the transmitter.

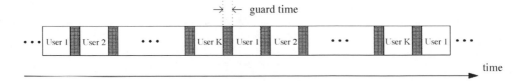

Figure 6-4 Slots for time-division multiple access

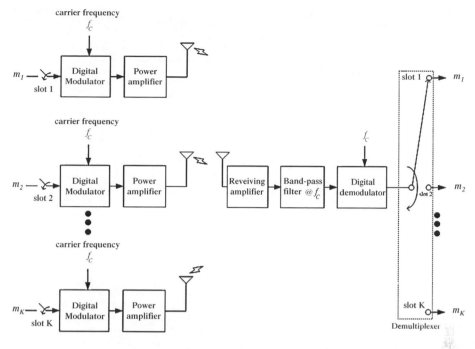

Figure 6-5 Example of a TDMA system

At the receiver, all transmitted signals are combined together at the receiving antenna. Then, a receiving amplifier is used to enlarge the received signal from the antenna, and a band pass filter is used to filter out unwanted noise. Since all users' signals are non-overlapping in the time domain, we can use a single demodulator to recover the messages from all users. Then, the demodulated messages are distributed to the corresponding users by a demultiplexer. The demultiplexer acts like a switch. If the demodulated output is obtained from slot 1, then the switch turns to the output port of user 1, and so on. Therefore, all users' messages can be recovered at the receiving end.

In a TDMA system, user k can transmit data only in the time slots assigned to user k. Therefore, each user's data are not transmitted continuously. Under this scenario, one may ask why the voice can be transmitted and received continuously in a TDMA system without any feeling of time-sharing. This problem is solved by partitioning the continuous voice signal into small segments. For example, for a four-user TDMA system, assume that each slot occupies 1 ms. Then each user can use one slot every 4 ms. The voice signal is then partitioned into segments of 4 ms each. Each segment is then converted (and compressed) into digital form. Assume that a total of B bits of voice data is produced for each segment of voice signal. The transmitter then transmits the B bits during each allowed 1 ms slot, as shown in Figure 6-6. The receiver receives each user's data at the corresponding time slot and reconstructs the voice signal in the following 4 ms. All reconstructed voice segments are concatenated in time, resulting in a continuous voice signal.

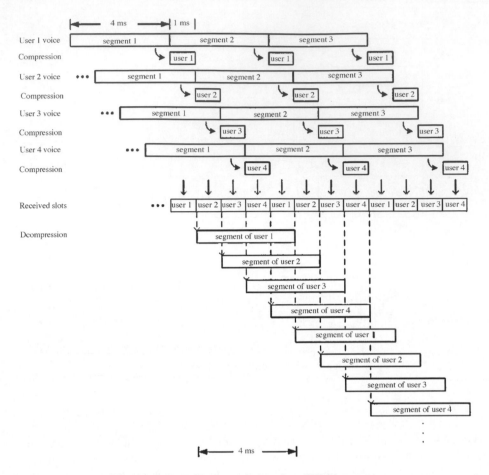

Figure 6-6   Voice transmissions in a TDMA system

Theoretically, TDMA can be implemented for analogue signals, but in practice it is much easier for TDMA to be implemented when the data are in digital form. There are many non-ideal effects in TDMA systems. For example, perfect time synchronization between individual users is not easy to achieve in a practical system. Therefore, the system must allow acceptable time-synchronization errors. Also, the carrier frequencies of different users may have slight differences, and the carrier phases of different users may be totally random. Therefore, at the receiver, a quick carrier recovering circuit is required so that the demodulator can switch and recover messages between different users.

Note that the bit rate as well as the bandwidth of a communication system will be increased if the TDMA system is used.

Having introduced the FDMA and TDMA systems, we can now introduce many systems using these technologies.

The GSM Mobile Phone System that uses Both FDMA and TDMA Technologies

For the GSM system, RF frequencies are in the range of 900, 1800 and 1900 MHz. That is, any company which provides a GSM mobile service uses an RF frequency around such frequencies. Each RF channel consists of around 124 subchannels, and each subchannel has a bandwidth around 0.2 MHz with eight TDMA systems. Thus, each GSM has a bandwidth around $124 \times 0.2 \approx 25$ MHz. Each carrier frequency is shared by eight users in the TDMA fashion. We can see that the total number of users is $124 \times 8 \approx 1000$. We may say that the GSM system allows around a maximum of 1000 users to access one base station. The GSM system uses Gaussian Minimal-shift Keying (GMSK), a technique which is quite similar to the FSK technique for digital modulation.

T1 to T4 Systems that use TDMA Technology

The T1 to T4 systems are transmission lines along which data are transmitted by using the baseband pulse transmission method. There is no carrier frequency involved, so these systems all use the TDMA technique. Its bit rate for T1 is 1.544 Mbps with 24 channels, each channel transmits data with 64 Kbps. The T2 system has 6.312 Mbps and is equal to four T1s, and thus 96 channels. The T3 system has 44.73 Mbps and is equivalent to seven T2s and has 672 channels. Finally, T4 has 274.17 Mbps, is equal to six T3s, and has 4032 channels.

Digital Enhanced Cordless Telecommunication (DECT) using Both FDMA and TDMA Technologies

In the home we often use cordless phones. Thus, some wireless communication technique must be used. Our present cordless phone system actually allows people in the house to talk to each other. Therefore it needs some multiplexing mechanism. The Digital Enhanced Cordless Telecommunication (DECT) products have today been widely accepted throughout the world for domestic, business, industrial and wireless local loop applications. In contrast to a digital mobile telephone system like GSM, cordless terminals generally transmit at lower power than mobile telephone systems, necessitating their use at a range of up to 100 m or so, compared to cell sizes of tens of kilometers for digital mobile telephone systems. The channel spacing of DECT is 1.728 MHz located in the 1880–1900 MHz band. DECT employs the TDMA technology with 24-timeslot per carrier frequency. Thus, a single DECT carrier can support multiple calls over a single RF transceiver. The modulation technique employed in the DECT is the Gaussian Frequency-shift Keying (GFSK) which is also a special form of the FSK introduced in Section 5.4.

One may wonder why we need such an elaborate system for a rather simple application. Of course, if only one telephone is used and we do not allow users in our homes to use these cordless phones to talk to each other, we really do not need such a sophisticated system. With the DECT system, which is similar to GSM, we allow family members to talk to one another although we seldom do that.

Digital Audio Broadcasting (DAB) Using both FDMA and TDMA Technologies

In Section 5.9, we introduced the DAB system. This system uses both FDMA and TDMA technologies because different broadcasting groups will use different frequency bands as OFDM is used for DAB. Besides, as noted before, TDMA is also used.

A Wireless Audio System

Communications technology allows wireless communication between a tuner, or a computer, and a speaker. The following are typical data for a wireless audio system. Its carrier frequency is 2.4 GHz and bandwidth 20 MHz. It employs DQPSK technology for digital modulation. It employs TDMA technology to serve three speakers. To prevent noise from corrupting the transmission, the system uses DSSS technology to spread spectrum. The spread spectrum mechanism will be discussed in Chapter 7.

6.3 Code-division Multiple Access (CDMA)

In the FDMA system, each user is given a unique frequency. In the TDMA system, each user is given a unique time slot. In the *code-division multiple access* (CDMA) method, each user is given a unique code to represent 1s and 0s. All of the signals thus overlap both in time and in frequency. The basic requirement is that the codes are orthogonal, which will be explained later.

6.3.1 The Two-user CDMA System

Consider the time-limited signals $s_1(t)$ and $s_2(t)$ over the time interval $(0, T)$ as shown in Figure 6-7. It is obvious that these two signals overlap in the time domain. The spectra of

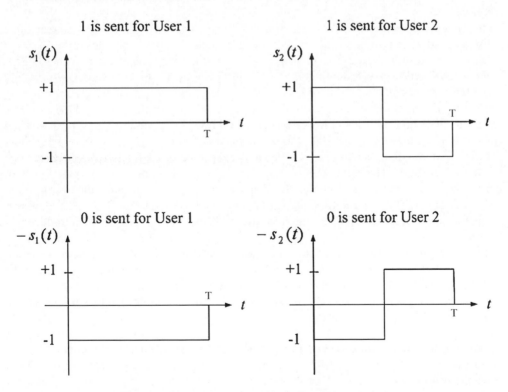

Figure 6-7   Orthogonal signals assigned to two users

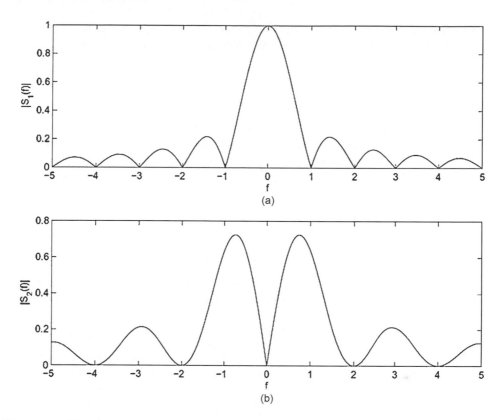

Figure 6-8 Magnitude of Fourier transforms of $s_1(t)$ and $s_2(t)$: (a) spectra of $s_1(t)$, and (b) spectra of $s_2(t)$

$s_1(t)$ and $s_2(t)$ also overlap in the frequency domain, as shown in Figure 6-8. Therefore, these two signals overlap in both domains, yet their inner product is zero:

$$\langle s_1(t), s_2(t) \rangle = \int_0^T s_1(t)s_2(t)dt = 0. \tag{6-1}$$

Signals $s_1(t)$ and $s_2(t)$ are mutually orthogonal over $(0, T]$ because of Equation (6-1).

A simple two-user multiple access communication system could be designed by letting users 1 and 2 modulate antipodally signals $s_1(t)$ and $s_2(t)$, respectively. *This means that user i transmits $s_i(t)$ in order to send 1 and $-s_i(t)$ in order to send 0.* Assume that the system is synchronous in the sense that the transmission rate is the same for both users and their signaling intervals are perfectly aligned. For the wireless environment, modulation techniques that transform the spectra of the transmitted signals to high-frequency bands are required. However, for simplicity, we consider only baseband CDMA systems; that is, CDMA systems without carrier modulation.

Let m_1 and m_2 be the binary messages in polar form (i.e. ± 1), transmitted by users 1 and 2, respectively. Then, the transmitted signals from both users can be expressed as

$$y_1(t) = m_1 s_1(t) \quad \text{for } 0 < t \le T$$
$$y_2(t) = m_2 s_2(t) \quad \text{for } 0 < t \le T. \tag{6-2}$$

Figure 6-9(a) and 6-9(b) show examples of $y_1(t)$ and $y_2(t)$ with the message sequence 10101101 for user 1 and 01101011 for user 2, where the signaling interval T is assumed to be one second. Assume that both transmitted signals are received perfectly. We can express the received signal as

$$r(t) = y_1(t) + y_2(t)$$
$$= m_1 s_1(t) + m_2 s_2(t). \tag{6-3}$$

The received signal is shown in Figure 6-9(c).

Figure 6-10 shows the model of a two-user CDMA system. We have assumed a noiseless case in this model. *At the receiver, the main job is to detect the corresponding messages m_1 and m_2 of Equation (6-3) for user 1 and user 2, respectively. As we discussed before, since $s_1(t)$ and $s_2(t)$ are orthogonal, $m_1(m_2)$ can be found easily by performing inner product*

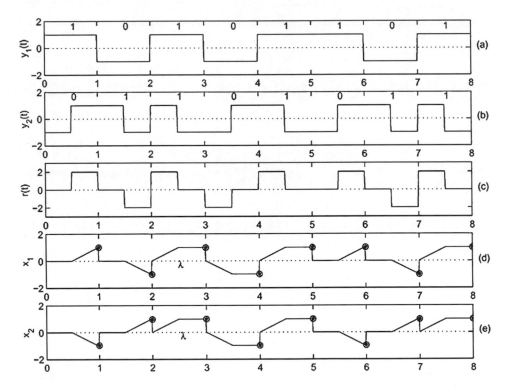

Figure 6-9 Example of signals of a two-user CDMA system: (a) $y_1(t)$, (b) $y_2(t)$, (c) $r(t)$, (d) $x_1(t)$, (e) $x_2(t)$

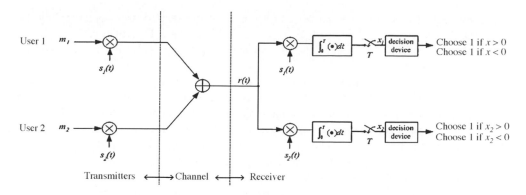

Figure 6-10   Model of a two-user CDMA system

between $r(t)$ and $s_1(t)(s_2(t))$. Thus, two correlators are used. For user 1, the sampled value x_1 from the first correlator can be expressed as

$$
\begin{aligned}
x_1 &= \int_0^T r(t)s_1(t)dt \\
&= \int_0^T (m_1 s_1(t) + m_2 s_2(t))s_1(t)dt \\
&= m_1 \int_0^T s_1^2(t)dt + m_2 \int_0^T s_2(t)s_1(t)dt.
\end{aligned}
\tag{6-4}
$$

Since $s_1(t)$ and $s_2(t)$ are orthogonal, the second term of the above equation is zero. Now, since $s_1^2(t) = 1$, the first term of the above equation equals $m_1 T$. Therefore we have

$$
x_1 = m_1 T.
\tag{6-5}
$$

The above equation shows that the correlator output of the first user contains only the transmitted message from user 1 and no interference from user 2. Similarly, for the second user, the sampled value x_2 from the second correlator can be expressed as

$$
\begin{aligned}
x_2 &= \int_0^T r(t)s_2(t)dt \\
&= \int_0^T (m_1 s_1(t) + m_2 s_2(t))s_2(t)dt \\
&= m_1 \int_0^T s_2(t)s_1(t)dt + m_2 \int_0^T s_2^2(t)dt.
\end{aligned}
\tag{6-6}
$$

Similarly we have

$$
x_2 = m_2 T.
\tag{6-7}
$$

The above equation shows that the correlator output of the second user contains only the transmitted message from user 2 and no interference from user 1. Figures 6-9(d) and 6-9(e) show the output signals of the correlators and the sampled points which are denoted by circles in the figure.

We have just seen a very simple example of a CDMA system. Users are assigned different 'signature waveforms' or 'codes'. Each transmitter sends its message by modulating its own signature waveform as in a single-user digital communication system. In the example above, the receiver need not concern itself with the fact that the signature waveforms overlap both in frequency and in time, because their orthogonality ensures that no interference from other users will appear in the output of one specific user's correlator.

Take another look at Equation (6-3). Pay attention to the fact that the two signals $s_1(t)$ and $s_2(t)$ are orthogonal. Because of this property, m_1 and m_2 can be determined easily. The reader is encouraged at this point to consult Sections 3.2 and 3.3.

6.3.2 The K-user CDMA System

For a general K-user CDMA system, each user is assigned a particular signature waveform which corresponds to a sequence of $+1$ and -1 of length N. Let $s_k = (s_{k0}, s_{k1}, \ldots, s_{k(N-1)})$ be the signature sequence of user k, where each s_{kj} equals $+1$ or -1 for $j = 0, 1, \ldots, N - 1$. Then the signature waveform of user k is defined by

$$s_k(t) = \sum_{j=0}^{N-1} s_{kj} p(t - jT_c), \tag{6-8}$$

where $T_c = T/N$ and $p(t)$ is a rectangular signal within $(0, T_c)$; that is, $p(t) = 1$ for $0 < t < T_c$ and $p(t) = 0$ elsewhere. The quantity T_c is called the *chip duration* and $1/T_c$ is called the *chip rate*. For example, a signature sequence $(+1,+1,-1,+1,-1, -1,+1,-1)$ of length 8 corresponds to the signature signal shown in Figure 6-11. It can be proved that the bandwidth of a CDMA signal has a broader spectrum than that of the original baseband signal. Therefore, the transmitted signal is expanded in the spectrum, and consequently a CDMA system is also called a *spread spectrum system*, which will be introduced in Chapter 7.

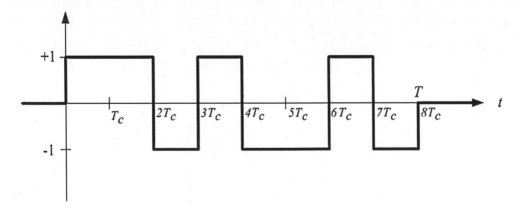

Figure 6-11   Signature waveform for the signature sequence $(+1, +1, -1, +1, -1, -1,+1, -1)$

Let us formally describe a CDMA system as follows. Let there be K users. In the CDMA system, user k uses his own code sequence $s_k(t)$ to represent binary symbol 1. User k sends $m_k s_k(t)$ where $m_k = \pm 1$. If 1(0) is sent, $m_k = 1(-1)$. We require that the K sequences be mutually orthogonal. These K sequences are mixed together to be sent as the message signal. Thus the sent signal is

$$s(t) = \sum_{k=1}^{K} m_k s_k(t). \tag{6-9}$$

Since the $s_k(t)$ are orthogonal, m_k can be easily detected by performing an inner product of $s(t)$ and $s_k(t)$. It can be easily seen that

$$\langle s(t), s_k(t) \rangle = NT_c m_k. \tag{6-10}$$

If the result of the inner product is positive, then we conclude that $m_k = +1$; otherwise, we conclude that $m_k = -1$.

The problem now is how to find a set of K orthogonal signature waveforms so that for $k \neq k'$,

$$\langle s_k(t), s_{k'}(t) \rangle = \int_0^T s_k(t) s_{k'}(t) dt = 0. \tag{6-11}$$

From Equation (6-8) we have

$$\langle s_k(t), s_{k'}(t) \rangle = \int_0^T \sum_{j=0}^{N-1} s_{kj} p(t - jT_c) \sum_{l=0}^{N-1} s_{k'l} p(t - lT_c) dt$$

$$= \sum_{j=0}^{N-1} \sum_{l=0}^{N-1} s_{kj} s_{k'l} \int_0^T p(t - jT_c) p(t - lT_c) dt. \tag{6-12}$$

Since $p(t)$ is a rectangular waveform over $(0, T_c)$, the integration of the product $p(T - jT_c)p(t - lT_c)$ equals zero if $j \neq l$ and equals T_c if $j = l$. Therefore, we have

$$\langle s_k(t), s_{k'}(t) \rangle = T_c \sum_{j=0}^{N-1} s_{kj} s_{k'j}. \tag{6-13}$$

From Equation (6-11), the inner product of $s_k(t)$ and $s_{k'}(t)$ is zero if and only if the inner product of the two vectors s_k and $s_{k'}$ is zero. For example, in the two-user system, $s_1 = (+1, +1)$ and $s_2 = (+1, -1)$. It is obvious that the inner product of s_1 and s_2 is zero and therefore $\langle s_1(t), s_2(t) \rangle = 0$.

For a set of N-dimensional signature sequences (vectors), there are at most N orthogonal vectors. We may generate a set of orthogonal vectors by Walsh functions. Walsh functions are generated by special square matrices called Hadamard matrices. These matrices contain one row of all $+1$, and the remaining rows each have equal numbers of $+1$ and -1. Walsh functions can be constructed for block length $N = 2^m$, where m is an integer. The $2^m \times 2^m$ matrix can be generated recursively as follows:

$$H_1 = [+1]$$

$$H_2 = \begin{bmatrix} +1 & +1 \\ +1 & -1 \end{bmatrix}$$

$$H_4 = \begin{bmatrix} +1 & +1 & +1 & +1 \\ +1 & -1 & +1 & -1 \\ +1 & +1 & -1 & -1 \\ +1 & -1 & -1 & +1 \end{bmatrix}$$

$$\vdots$$

$$H_{2^m} = \begin{bmatrix} H_{2^{m-1}} & H_{2^{m-1}} \\ H_{2^{m-1}} & -H_{2^{m-1}} \end{bmatrix}.$$

It can be shown that the inner product of different rows from H_{2^m} equals zero. Therefore, the signature sequence s_k for user k can be assigned as the kth row of H_N with $N = K = 2^m$. This method guarantees that two arbitrary different signature signals are orthogonal.

For example, if $K = N = 8$, the Hadamand matrix H_8 equals

$$H_8 = \begin{bmatrix} +1 & +1 & +1 & +1 & +1 & +1 & +1 & +1 \\ +1 & -1 & +1 & -1 & +1 & -1 & +1 & -1 \\ +1 & +1 & -1 & -1 & +1 & +1 & -1 & -1 \\ +1 & -1 & -1 & +1 & +1 & -1 & -1 & +1 \\ +1 & +1 & +1 & +1 & -1 & -1 & -1 & -1 \\ +1 & -1 & +1 & -1 & -1 & +1 & -1 & +1 \\ +1 & +1 & -1 & -1 & -1 & -1 & +1 & +1 \\ +1 & -1 & -1 & +1 & -1 & +1 & +1 & -1 \end{bmatrix}.$$

The resulting signature waveforms $s_1(t), s_2(t), \ldots, s_8(t)$ are depicted in Figure 6-12. For this set of signature waveforms, there are eight users.

Let us now use an example to explain why the K-user CDMA system works in a simple vector representation. Suppose we have eight users. Each user will use one of the rows to represent 1. That is, when user k sends 1, it sends s_k, the kth row of H_8. When user k sends 0, it sends $-s_k$, the negation of the kth row of H_8. For instance, user 2 will send $s_2 = [+1, -1, +1, -1, +1, -1, +1, -1]$ if he intends to send 1, and user 3 will send $-s_3 = [-1, -1, +1, +1, -1, -1, +1, +1]$ if he intends to send 0.

Consider the case where users 1, 2 and 3 send 1, 0 and 0, respectively; that is, $m_1 = +1$, $m_2 = -1$ and $m_3 = -1$. The following signals will be sent:

$$y_1 = m_1 s_1 = +s_1 = [+1, +1, +1, +1, +1, +1, +1, +1]$$
$$y_2 = m_2 s_2 = -s_2 = [-1, +1, -1, +1, -1, +1, -1, +1]$$
$$y_3 = m_3 s_3 = -s_3 = [-1, -1, +1, +1, -1, -1, +1, +1].$$

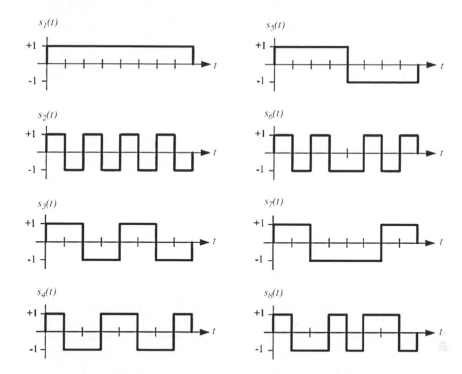

Figure 6-12 Signature signals generated from H_8

The received signal will be

$$z = y_1 + y_2 + y_3 = [-1, +1, +1, +3, -1, +1, +1, +3].$$

The transmitted signals, namely y_1, y_2 and y_3, can be recovered by using rows 1, 2 and 3 of H_8. Note that these vectors represent 1s for users 1, 2 and 3.

We now apply the inner product to recover y_1, y_2 and y_3. We compute $z \cdot s_1$, $z \cdot s_2$ and $z \cdot s_3$ as follows:

$$
\begin{aligned}
x_1 &= z \cdot s_1 \\
&= [-1, +1, +1, +3, -1, +1, +1, +3] \cdot [+1, +1, +1, +1, +1, +1, +1, +1] \\
&= -1 + 1 + 1 + 3 - 1 + 1 + 1 + 3 \\
&= 8.
\end{aligned}
$$

So, since $x_1 > 0$, we set $m_1 = +1$. Next:

$$
\begin{aligned}
x_2 &= z \cdot s_2 \\
&= [-1, +1, +1, +3, -1, +1, +1, +3] \cdot [+1, -1, +1, -1, +1, -1, +1, -1] \\
&= -1 - 1 + 1 - 3 - 1 - 1 + 1 - 3 \\
&= -8.
\end{aligned}
$$

Since $x_2 < 0$, we set $m_2 = -1$. Finally:

$$x_3 = z \cdot s_3$$
$$= [-1, +1, +1, +3, -1, +1, +1, +3] \cdot [+1, +1, -1, -1, +1, +1, -1, -1]$$
$$= -1 + 1 - 1 - 3 - 1 + 1 - 1 - 3$$
$$= -8$$

Again, since $x_3 < 0$, we set $m_3 = -1$. Thus we can see that all of the original sent signals are correctly identified.

In general, the sent signal is $z = \sum m_k s_k$, where $m_k = +1$ or -1. After z is received, we compute $x_k = z \cdot s_k$. If $x_k > 0$ we set $m_k = +1$, and if $x_k < 0$ we set $m_k = -1$. That we can do so is due to the fact that the s_k are orthogonal to one another.

One may ask whether the receiver can determine the other users, namely users 4 to 8 have sent nothing. To do this, the receiver may simply perform appropriate inner products. For instance, we can easily see that $z \cdot s_4 = 0$, which indicates that user 4 has sent nothing.

The bandwidth of the CDMA is larger because each symbol consists of a number of bits and each bit must be of smaller pulse length. As we indicated before, a small pulse length will give us a larger bandwidth. What is the advantage of having a larger bandwidth? One of the advantages is that it helps us solve 'the multipath problem'. Note that the signal, after it leaves the transmitter, may go into different directions and reflections may occur. Thus, one signal becomes several signals and they will all reach the receiver. This problem is called the multipath problem. A larger bandwidth will make us easier to overcome this difficulty. The CDMA technology is used in the third-generation mobile systems because, when we use a mobile phone, we may be moving and the multipath problem can be a serious one. In such an environment, CDMA is of course quite desirable.

6.4 Carrier-sense Multiple Access (CSMA)

Carrier-sense multiple access (CSMA) is widely used in a local-area network (LAN) where several host computers are connected. All of the computers will talk to one another. One such LAN is the Ethernet. Essentially, an Ethernet consists of a cable as shown in Figure 6-13.

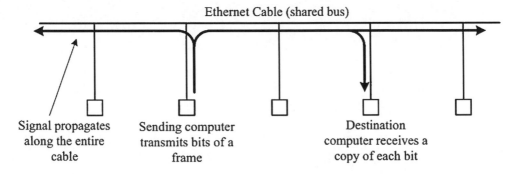

Figure 6-13   The Ethernet system

Ethernet employs a bus. Naturally, only one user can have access to it. It is important to note that Ethernet is a computer network. Information flowing through it is data, not, say voice. Thus there is no requirement that the information has to be received by someone in real time. Let us assume that there is only one computer, say computer A, which is sending data to computer B. Data are in digital form and prepared into frames, or packets. Each frame will contain some control information. For instance, we have to let the network know who is sending this frame, who is supposed to receive the frame and so on. As the frame propagates from computer A to the Ethernet bus, it fully occupies the bus. Thus all computers connected to the bus can receive the frame. Since the control information of the frame tells us that computer B is to receive the frame, all other computers ignore it and only computer B receives it.

Since the Ethernet allows a sender to fully occupy it, each computer must check whether anyone is occupying the bus. The computer has to have the ability to sense whether there is any signal transmitted over the bus. This is why this kind of checking whether there is a signal occupying the bus is called 'carrier sense'.

Since a bus allows only one user to occupy it, when two computers simultaneously send signals to the bus, collision occurs and this must be resolved. The Ethernet enables each sender to be able to monitor whether a collision occurs or not. If a sender senses a collision, it backs off immediately. That is, it stops sending signals to the bus. It does not send the signal back right away. Instead, it waits for a certain period of time and transmits again. A random number generator determines the length of this period of time. Thus, if two senders simultaneously back off, the probability that they will send signals back at the same time is very small. This kind of CSMA system is called *carrier-sense multiple access with collision detection* (CSMA/CD).

Another way of solving the collision problem in a CSMA system is to avoid collisions at the very beginning. The system requires each user to send an exceedingly short message signal to the bus to *reserve* the system. If a collision of this reservation occurs, it is solved by the method discussed in the above paragraph. Because these reservation messages are very short, the overhead caused by resolving their collision is quite slight. After a sender successfully reserves the bus, it can send signals and only it can send signals. This kind of system is called a *carrier-sense multiple-access with collision avoidance* (CSMA/CA).

We may say that CSMA is a concept. It can be used in a wireless environment as well. The IEEE 802.11a WLAN system uses this technology.

6.5 The Multiplexing Transmission Problem

We should note that a local area network not only transmits signals inside the network. It is also necessary for it to receive a signal and send it to some other network. In general, for multiplexing communications, there are two problems:

- The first is the multiplexing receiving problem. Signals must be clearly identified and received.
- The second is the multiplexing transmitting problem. That is, after signals are received, they must be put together and sent to their destinations correctly.

A typical case is the telephone system. A central office or a base station in the mobile phone case receives many calls. They must be transmitted to correct destinations. In this section, we shall discuss this multiplexing transmission problem.

There are essentially two methods to transmit multiplexing information. The first one is the *circuit switching system*. Telephone traffic is now mostly handled by this switching system.

After we dial the telephone number of the person whom we want to talk to, the system first checks whether the phone of the other person is on hook or not. If yes, it sends a signal to start the phone ringing. At the same time, a fixed route, consisting of trunks, is found for our call. In other words, whenever we want to talk, our voice will be sent to the other phone through this route. Similarly, if the other person wants to talk, his voice will also be sent to us through this route.

Note that this route is not used by one caller only. To fully utilize a trunk, the voices of many callers must be sent through it at the same time. The TDMA is usually used to achieve this goal.

A telephone system must be a real-time system because your voice must be sent to the caller with delay so small that the person you called will feel that your voice is continuous and cannot detect any distance between you and him. The circuit switching system uses a fixed route. This guarantees that your voice will not be delayed because the route is always available to you. While this is an advantage, it is also a disadvantage because the callers may pause for a while. While the phones are silent, the route will not be fully utilized. Since telephone lines are increasingly being used to transmit data, in addition to voice, the packet switching concept was invented.

In the circuit switching system, information is divided into fixed size packets. In the packet switching system, information is also divided into packets, but they may have different sizes. At the head of each packet, more control information is needed. The major difference between the circuit switching and the packet switching systems is that *the packet switching is more software controlled*. Supposedly, each node in the network knows the whole situation of the network. Thus, a packet will not be sent through a fixed route; rather it will be sent through an optimal route. This implies that the system has to use some decision rule to decide the route dynamically, and hence delay naturally may occur in a packet switching system. There is no 100% pure packet switching system; present-day systems are all variations of the packet switching system concept.

Further Reading

- For more details about multiuser communications, see [V98].
- For more details about code division multiple access, see [Z04].

Exercises

6.1   Consider the superheterodyne system. Is this an FDMA system?

6.2   Consider the FSK system. Is this an FDMA system? For one receiver, how many band-pass filters does it need?

6.3   Suppose we have to implement two receivers. Receiver A receives data with higher bit rate and receiver B receives data with lower bit rate. Both use the FDMA technique. Which one is easier to design? Explain.

6.4   The OFDM system obviously has to implement many band-pass filters. Does this present a hard problem? Why?

6.5   For the CDMA system whose code is shown in Figure 6-12, assume that the received vector is $[-1,+1,+1,-1,+3,+1,-3,-1]$. Determine the code sent by each user. Remember that a user may have sent nothing.

6.6   Consider a system whose bandwidth is fixed and only one carrier frequency is used. Suppose each subchannel is used to transmit high-quality music. In such a case, TDMA must be used. Can you have a large number of subchannels without downgrading the quality of music transmitted?

6.7   Suppose a telephone system uses both Internet and CSMA technology. Can it guarantee good quality of voice transmission?

7

Spread-spectrum Communications

In a communications system, it is usually important to utilize the bandwidth efficiently; that is, we prefer the bandwidth to be small. However, it is sometimes necessary to sacrifice this bandwidth criterion to meet other design objectives. For example, in a secure wireless system, the transmitted signal must not be easy to be detected or recognize by unauthorized listeners. In such systems, security is more important than the efficient use of bandwidth.

Spread-spectrum communication was originally developed for military applications. Essentially, it widens the bandwidth. One major concern of the design objective of such a system is the resistance to interference (jamming). However, there are civilian applications that employ the spread-spectrum technique. For example, spread-spectrum was adopted in the IEEE 802.11 wireless local-area network (WLAN) standard to provide multipath rejection in a wireless data link environment. (The multipath problem is beyond the scope of this book.) Another application is in multiple-access communications in which a number of independent users are required to transmit information via a common channel. We shall see that the spread-spectrum technique can be used in the code-division multiple-access (CDMA) system which was introduced in Chapter 7.

7.1 The Basic Concept of Spread-spectrum

Consider Figure 7-1(a) which contains a pulse. The Fourier transform spectrum of this pulse is shown in 7-1(b), as discussed in Section 4.3. Now, suppose we reduce the pulse width as in 7-1(c); then the corresponding Fourier transform spectrum will be widened, as shown in 7-1(d).

As the pulse width is reduced, the Fourier transform spectrum is spread out, in some sense. What is the significance of this spreading? Note that the signal now contains more frequency components. There are at least two advantages of this spreading of spectrum:

- When the bandwidth of a spectrum is widened, it is relatively more difficult for an intruder to understand the signal as he will have to have a receiver which can detect a wide spectrum of signals.
- If the bandwidth of a spectrum is narrow, any noise will be significant. Thus the signal will be easily disturbed by the noise. On the other hand, if the spectrum is spread out, the noise will not have a significant effect on the signal unless it occupies itself a rather wide spectrum.

Communications Engineering. R.C.T. Lee, Mao-Ching Chiu and Jung-Shan Lin
© 2007 John Wiley & Sons (Asia) Pte Ltd

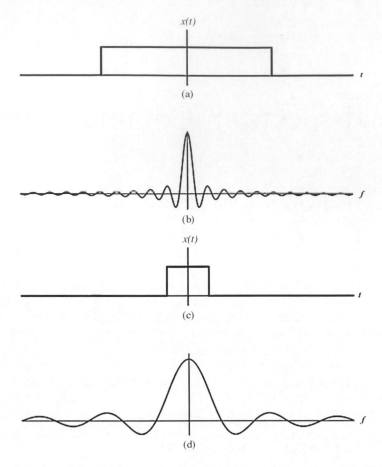

Figure 7-1 Changing of the Fourier transform spectrum with respect to the pulse width: (a) a wide pulse; (b) Fourier transform of the pulse in (a); (c) a narrow pulse; (d) Fourier transform of the pulse in (c)

Let there be two signals $x(t)$ and $y(t)$. Let their Fourier transforms be denoted as $X(f)$ and $Y(f)$, respectively. To simplify our discussion, let us assume that $X(f)$ and $Y(f)$ are square waves as shown in Figures 7-2(a) and 7-2(b), respectively. We assume that the areas under $X(f)$ and $Y(f)$ are $P_1a_1 = 1$ and $P_2a_2 = 1$. In addition, the bandwidth of $X(f)$ is wider than that of $Y(f)$, so $a_1 > a_2$. For simplicity, we may call $X(f)$ a wide-band signal and $Y(f)$ a narrow-band signal.

As pointed out in Chapter 3, the Fourier transform of the signal $x(t)y(t)$ is the convolution of $X(f)$ and $Y(f)$. To find the convolution, we can imagine $Y(f)$ to be fixed in space and $X(f)$ to be moved from minus infinity to the right. The value of the convolution will be 0 until $Y(f)$ hits $X(f)$. Thus $X(f)*Y(f)$ starts to be nonzero at $f = -(1/2)(a_1 + a_2)$, as shown in Figure 7-2(c). It will continue to increase until the right leading edge of $X(f)$ coincides with that of $Y(f)$, as shown in Figure 7-2(d). That is, at $f = -(1/2)(a_1 - a_2)$, the convolution reaches its peak which is $P_1P_2a_2 = P_1$. The entire function of the convolution is shown in Figure 7-2(e).

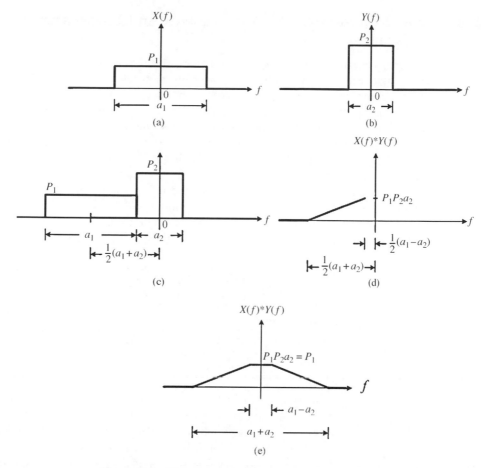

Figure 7-2   The convolution of two square-wave spectra

From Figure 7-2(e), we can see that the convolution of two spectra has a tendency of flattening the spectrals. That is, the spectrum of $x(t)y(t)$ is wider than both $X(f)$ and $Y(f)$. *This means that by multiplying $x(t)$ to $y(t)$, we can effectively spread the spectrum of $y(t)$ if the spectrum of $X(f)$ is wide enough. This is quite significant and is a basis of one of the spread-spectrum systems. The reader can see that $x(t)$ can be seen as a modulating signal. The purpose of this modulation is not to lift the frequency, rather to spread the spectrum.*

Since $a_1 > a_2$, we have $P_1 < P_2$. From Figure 7-2(e), we can see that the multiplication of the narrow-band signal $y(t)$ and the wide band signal $x(t)$ has an effect of reducing the magnitude of the spectrum of $y(t)$ from P_2 to $P_1 P_2 a_2 = P_1$. This property is also significant as we shall see later.

One problem of using the Fourier transform is that the coefficient of a Fourier transform can be complex. In the discussion of spread-spectrum systems, we shall use the power spectrum concept which was introduced in Chapter 5.

7.2 Baseband Transmission for Direct-sequence Spread-spectrum (DSSS) Communications

One important characteristic of spread-spectrum communication is that it can provide protection against interfering signals with finite power by purposely making the information-bearing signal occupy a bandwidth far in excess of the minimum necessary to transmit it. In addition, the transmitted signal has the feature that it is a noise-like signal with background noise hidden in it. The transmitted signal is thus safe to propagate through the channel without being detected by unauthorized listeners. In this section we shall introduce the direct-sequence spread-spectrum (DSSS) method. To illustrate the idea of DSSS, we first introduce baseband transmission of a DSSS signal.

One method of spreading the bandwidth of the baseband information signal involves the use of modulation. Let $\{m_k\}$ be a binary data sequence with $m_k \in \{+1, -1\}$. Let T_b be the bit signaling interval for transmission of $\{m_k\}$. Then the data rate can be expressed as $1/T_b$ bits/s. The baseband pulse signal of the information sequence can be expressed as

$$m(t) = \sum_k m_k p(t - kT_b), \tag{7-1}$$

where $p(t)$ is the pulse shape signal defined as

$$p(t) = \begin{cases} 1 & 0 < t < T_b \\ 0 & \text{otherwise.} \end{cases} \tag{7-2}$$

For example, let the information sequence $\{m_k\}$ be $\{+1, -1, -1, +1\}$. The baseband pulse signal $m(t)$ with $T_b = 1\,\text{s}$ for the information sequence $\{+1, -1, -1, +1\}$ is depicted in Figure 7-3(a). We will refer to $m(t)$ as the information-bearing signal. In the frequency domain, the power spectral density of $m(t)$ is shown in Figure 7-4(a). Because of the properties of power spectral density discussed in the previous section, if the signaling interval T_b becomes smaller, the wider the bandwidth of the power spectral density may be. The power spectral density of $m(t)$ concentrates over the range $[-1/T_b, 1/T_b]$, which is $[-1, +1]$ in this case since $T_b = 1\,\text{s}$. Therefore, we may define the bandwidth of $m(t)$ as $2/T_b$.

Now, to spread the bandwidth of the transmitted signal, we may multiply $m(t)$ with a signal $c(t)$ that has a power spectral density wider than that of $m(t)$. From Fourier transform theory, multiplication of two signals produces a signal whose spectrum equals the convolution of the spectra of the two signals as informally shown in Section 7.1. Thus if the message signal $m(t)$ is of narrow band and the signal $c(t)$ is of wide band, the product signal $m(t)c(t)$ will have a spectrum that is nearly the same as the wide-band signal $c(t)$.

Now the problem is how to generate $c(t)$. The generation of $c(t)$ relies on a man-made noise-like binary sequence $\{c_k\}$ called a *pseudo noise sequence* (PN sequence). A detailed description of the PN sequence will be given in Section 7.4. Right now, we only need to remember that $\{c_k\}$ is a random-like sequence with $c_k \in \{+1, -1\}$ and is known by both of the transmitter and the receiver. Mathematically, $c(t)$ can be expressed as

$$c(t) = \sum_k c_k q(t - kT_c), \tag{7-3}$$

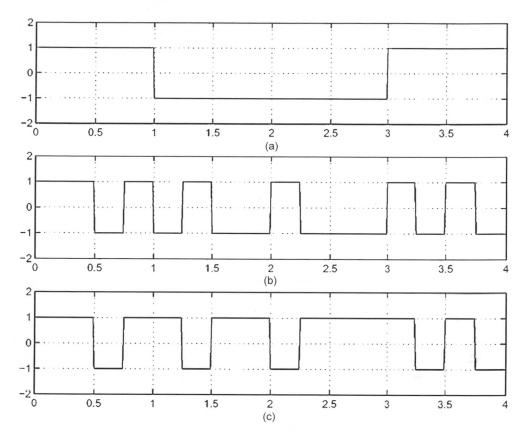

Figure 7-3 Signals in the DSSS technology: (a) waveform of baseband signal $m(t)$; (b) PN signal $c(t)$; (c) product signal $y(t) = m(t)c(t)$

where T_c is called the **chip interval** and $q(t)$ is a pulse signal defined as

$$q(t) = \begin{cases} 1 & 0 < t < T_c \\ 0 & \text{otherwise.} \end{cases} \tag{7-4}$$

To make the spectrum of $c(t)$ wider than that of $m(t)$, the chip interval T_c must be smaller than the signaling interval T_b. In most cases, the chip interval is set to be $T_c = T_b/N$, where N is a positive integer greater than unity. It is obvious that the bandwidth of $c(t)$ is $2/T_c = 2N/T_b$ which is N times wider than that of $m(t)$. Figure 7-3(b) shows a PN signal $c(t)$ with the chip rate $T_c = T_b/4$ and PN sequence c_k being $\{+1, +1, -1, +1, -1, +1, -1, -1, +1, -1, -1, -1, +1, -1, +1, -1\}$. Figure 7-4(b) shows the power spectral density of $c(t)$. It is obvious that the bandwidth of $c(t)$ is $2/T_c = 8/T_b$ which is four times wider than that of $m(t)$.

By multiplying the information-bearing signal $m(t)$ by the PN signal $c(t)$, each information bit is divided into a number of smaller time intervals. These time intervals are

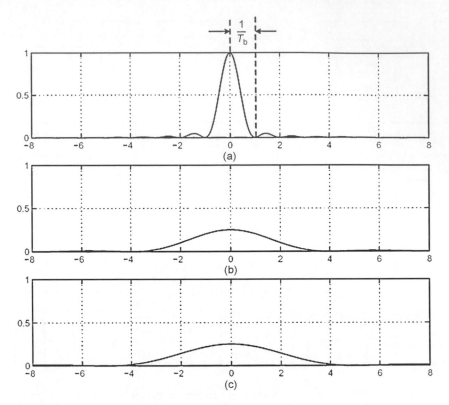

Figure 7-4   Power spectral densities of different signals: (a) of the baseband signal $m(t)$; (b) of the PN signal $c(t)$; (c) of the product signal $y(t) = m(t)c(t)$

commonly referred to simply as ***chips***. For baseband transmission, the resulting signal $y(t)$, which represents the transmitted signal, is expressed as

$$y(t) = m(t)c(t). \tag{7-5}$$

An example of the waveform of the resulting signal $y(t)$ is given in 7.3(c), and the power spectral density of $y(t)$ is given in 7-4(c). ***Comparing Figures 7-4(a) and 7-4(c), we see that the bandwidth of the transmitted signal $y(t)$ is much wider than that of the original information-bearing signal $m(t)$.*** Besides, in the frequency domain, as pointed out in Section 7.1, the magnitude of the power spectral density of the wide-band signal $y(t)$ will be smaller than that of the narrow-band signal $m(t)$ because of the convolution mechanism. ***For an outsider to fully comprehend $y(t)$, he needs a receiver system which covers a wider range of frequencies, for each frequency he needs to be able to detect a signal with a rather small amplitude, and he needs to know $c(t)$. Thus, the spread-spectrum system can be viewed as an encryption system which prevents an outsider comprehending the signal transmitted***.

Modulation and demodulation of the spread-spectrum system introduced above is shown in Figure 7-5.

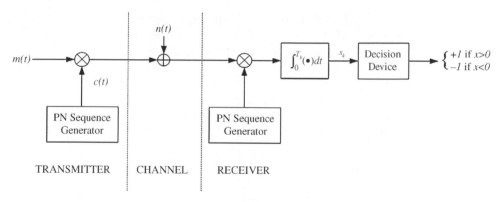

Figure 7-5   A spread-spectrum modulation and demodulation system

On the left of Figure 7-5 is the DSSS modulator that performs the multiplication, where *the PN signal $c(t)$ is generated by a PN sequence generator*. The received signal $r(t)$ consists of the transmitted signal $y(t)$ and an additive interference signal denoted by $n(t)$ as shown in Figure 7-5. Hence:

$$r(t) = y(t) + n(t)$$
$$= m(t)c(t) + n(t), \tag{7-6}$$

where the noise term $n(t)$ *may be produced by other sources outside of the system which intentionally wish to break down the communication system or may be unintentionally generated from the circuit itself*. To recover the original message signal $m(t)$, the received signal $r(t)$ is passed to a demodulator consisting of a multiplier followed by an integrator, and a decision device as shown in the right part of Figure 7-5. The multiplier is supplied with a locally generated PN signal $c(t)$ which is an exact replica of that used in the transmitter. The process of multiplying $c(t)$ to the received signal is called 'despreading'. This requires that the PN sequence in the receiver be lined up exactly with that in the transmitter. The multiplier output is therefore given by

$$z(t) = r(t)c(t)$$
$$= m(t)c^2(t) + n(t)c(t). \tag{7-7}$$

The above equation shows that the data signal $m(t)$ is multiplied twice by the PN signal $c(t)$. Since, from Equation (7-3), $c(t)$ is a bipolar baseband signal (i.e. $c(t) \in \{+1, -1\}$ for all t), we have $c^2(t) = 1$ for all t. Accordingly, we have

$$z(t) = m(t) + n(t)c(t). \tag{7-8}$$

From Equation (7-8), the information-bearing signal $m(t)$ is reproduced at the multiplier output in the receiver, except the additive term $n(t)c(t)$ which represents the effect of interference.

Our main concern is whether this term $n(t)c(t)$ will have a great influence on $m(t)$ or not. Note that the interference term $n(t)$ is multiplied by the locally generated PN signal $c(t)$. This means that the spreading code $c(t)$ will affect the interference signal $n(t)$ just as it did the original signal $m(t)$ at the transmitter end. We observe that the information-bearing signal $m(t)$ is of narrow band, while the term $n(t)c(t)$ is of wide band since $c(t)$ is of wide band. To explain how the effect of $n(t)$ is reduced, we assume that $n(t) = A\cos(2\pi f_j t)$. The power spectral density of $n(t)$ is $P_n(f) = (A^2/4)(\delta(f - f_j) + \delta(f + f_j))$ which represents an extremely large power allocated at frequencies f_j and $-f_j$. The power of $n(t)$ can be found as

$$
\begin{aligned}
P_n &= \lim_{T \to \infty} \frac{1}{T} \int_{-T/2}^{T/2} |n(t)|^2 dt \\
&= \lim_{T \to \infty} \frac{1}{T} \int_{-T/2}^{T/2} |A\cos(2\pi f_j t)|^2 dt \\
&= \lim_{T \to \infty} \frac{1}{T} \int_{-T/2}^{T/2} \frac{A^2}{2} dt \\
&= \frac{A^2}{2}.
\end{aligned}
\tag{7-9}
$$

Now define $n'(t) = n(t)c(t)$ and define $P_{n'}(f)$ and $P_c(f)$ as the power spectral densities of $n'(t)$ and $c(t)$, respectively. It is obvious that the power spectral density of $n'(t)$ is

$$
P_{n'}(f) = (A^2/4)P_c(f - f_j) + (A^2/4)P_c(f + f_j)
\tag{7-10}
$$

by the convolution theorem. This means that the power spectral density of $n'(t)$ is shifted towards frequencies f_j and $-f_j$. The bandwidth of the $P_{n'}(f)$ is now larger than that of $P_c(f)$ which is $2N/T_b$ and is much wider than that of $P_n(f)$. Because the bandwidth is now enlarged, we may expect that signal $n'(t) = n(t)c(t)$ would interfere with the signal $m(t)$. *In Section 7.2 we saw that the convolution of two spectra will result in a spectrum which has smaller amplitude. Therefore, we may expect the magnitude of the power spectral density of $n'(t) = n(t)c(t)$ to be smaller than that of $n(t)$. This will be formally proved as follows.*

For simplicity, assume that the power spectral density of $c(t)$ is a constant W over $(-N/T_b, N/T_b)$ and zero otherwise:

$$
P_c(f) = \begin{cases} W & -N/T_b < f < N/T_b \\ 0 & \text{otherwise.} \end{cases}
\tag{7-11}
$$

Since $c(t) \in \{+1, -1\}$, the average power of $c(t)$ is 1. We have

$$
P_c = 1 = \int_{-N/T_b}^{N/T_b} W df = 2WN/T_b,
\tag{7-12}
$$

which indicates that $W = T_b/(2N)$. From Equation (7-10), since $P_c(f - f_j)$ and $P_c(f + f_j)$ overlap and $P_c(f) = W$ for $-N/T_b < f < N/T_b$, the maximum magnitude of $P_{n'}(f) = (A^2/4)P_c(f - f_j) + (A^2/4)P_c(f + f_j)$ is therefore equal to $2(A^2W/4) = A^2W/2$

$= A^2 T_b/(4N)$, which is much less than $(A^2/4)\delta(f - f_j)$ and $(A^2/4)\delta(f + f_j)$. The effect of $n(t)$ is then significantly reduced by despreading.

The above discussion shows that the effect of interference signal $n(t)$ on $m(t)$ is not that significant because it is multiplied by $c(t)$. By applying the output of the multiplier to a low-pass filter with a bandwidth just large enough for recovering the information-bearing signal $m(t)$, most of the power in the component $n(t)c(t)$ is filtered out.

Figure 7-6 illustrates the spectra of related signals and how the interference effect is reduced. Figure 7-6(a) depicts the power spectrum of the information-bearing signal $m(t)$. The spectrum of the received signal is shown in 7-6(b) where the spectrum of the spreading signal $m(t)c(t)$ is depicted. In addition, we assume that there is a strong narrow-band interference $n(t)$ whose power spectrum concentrates over a narrow range of frequency as depicted in 7.6(b). At the receiver, the despreading signal $z(t)$ given in Equation (7-8) contains two components. The first component is the spectrum of the recovered signal $m(t)$ and the second component is that of the interference $n(t)c(t)$ as given in Figure 7-6(c). *Although the interference signal $n(t)$ becomes a wide-band signal $n(t)c(t)$, the magnitude*

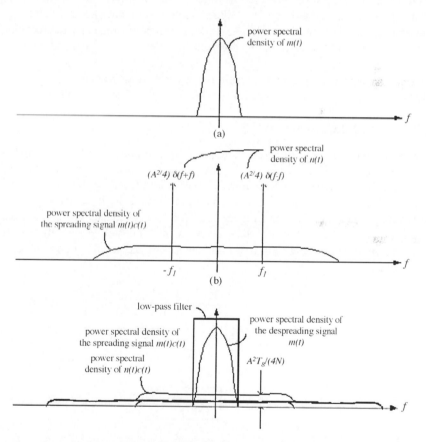

Figure 7-6 Power density spectra of $m(t), n(t)$ and $m(t)c(t)$ and the despreading signal: (a) of the baseband signal $m(t)$; (b) of the spreading signal $m(t)c(t)$ and a narrow band interference $n(t)$; (c) of the despreading signal $m(t)$ and the interference $n(t)c(t)$

is dramatically reduced over the band of interest because of the multiplication of $c(t)$ by $n(t)$. This is the reason why the effect of the interference is reduced by spreading and despreading techniques. The above discussion shows that the spread-spectrum system is more immune to outside interference which is the second characteristics of the system.

In summary, there are two important characteristics of the spread spectrum system: (1) it makes it more difficult for an unauthorized listener to comprehend the transmitted signal; and (2) it is more immune to outside interference.

We indicated before that the received signal will be passed through a low-pass filter to filter out the high-frequency components and only keep the low-frequency components within the message signal $m(t)$. This can be done through an integration and dump circuit, called a correlator in Chapter 5 and shown in Figure 7-5. The integration evaluates the area under the signal produced at the multiplication output. The integration is performed for the bit interval of the message signal $m(t)$, namely T_b. As we repeatedly pointed out, this bit interval corresponds to the low-frequency components of $m(t)$. Note that, actually, T_b is relatively wide as compared with the chip interval of $c(t)$. Accordingly, it corresponds to low frequencies and the integration of $n(t)c(t)$ over the interval T_b will be very small. Thus the integrator can be considered as a low-pass filter. We note here that the outside interference is not significant because of the dispreading. Finally, a decision is made by the receiver. If x is greater than the threshold of zero, the receiver decides that $+1$ was sent; otherwise, -1 was sent.

In Chapter 6 we gave a typical specification of a wireless audio system. It uses the DSSS technology to prevent noise from interfering with the transmission of music.

7.3 BPSK Modulation for DSSS

In the previous section we considered baseband transmission for a DSSS signal, without carrier modulation. For wireless communication the signal must be transmitted as RF (radio frequency), so a carrier frequency level must be available. We may incorporate binary phase-shift keying (BPSK), or any other digital modulation which produces RF signals, into the transmitter and receiver, as shown in Figure 7-7. The binary data sequence m_k is passed by two stages of modulation. The first stage consists of a multiplier with the information-bearing signal $m(t)$ and the PN signal $c(t)$ as inputs. The second stage consists of a BPSK

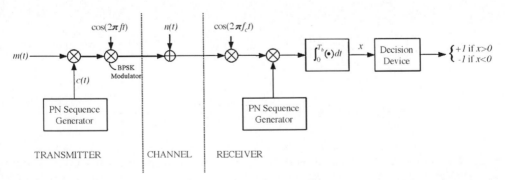

Figure 7-7   Transmitter and receiver for direct-sequence spread coherent phase-shift keying

modulator. The transmitted $s(t)$ is therefore a ***direct-sequence spread binary phase-shift-keyed*** **(DS/BPSK)** signal.

The transmitted signal is thus obtained by multiplying the baseband DSSS signal $y(t) = m(t)c(t)$ by a carrier signal $\cos(2\pi f_c t)$ with f_c greater than $2/T_c$. The BPSK signal can now be expressed as

$$s(t) = m(t)c(t)\cos(2\pi f_c t). \tag{7-13}$$

Figure 7-8 illustrates the waveforms for all the related signals. Figure 7-8(a) gives an example of an information-bearing signal $m(t)$. The PN signal $c(t)$ is in 7-8(b). The transmitted BPSK signal $m(t)c(t)\cos(2\pi f_c t)$ is in 7-8(c), where a carrier waveform appears in the signal.

The receiver, shown in the right part of Figure 7-7, performs two stages of demodulation. In the first stage, the received signal is multiplied by a locally generated carrier waveform $\cos(2\pi f_c t)$. This is a typical down-conversion operation. Thus the signal now becomes $m(t)c(t)\cos^2(2\pi f_c t)$ as shown in Figure 7-8(d). The second stage of demodulation performs spectrum despreading by multiplying the input signal by a locally generated replica of the

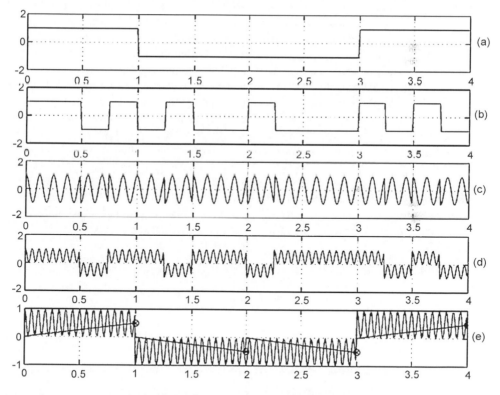

Figure 7-8 BPSK modulation for DSSS: (a) information-bearing signal $m(t)$; (b) PN signal $c(t)$; (c) transmitted signal $m(t)c(t)\cos(2\pi f_c t)$; (d) $m(t)c(t)\cos^2(2\pi f_c t)$, the output signal from the first stage of the demodulator; (e) $m(t)c^2(t)\cos^2(2\pi f_c t) = m(t)\cos^2(2\pi f_c t)$, the despread signal, integrated signal, and the sampled points

PN signal $c(t)$. The signal now becomes $m(t)c^2(t)\cos^2(2\pi f_c t) = m(t)\cos^2(2\pi f_c t)$ as shown in Figure 7-8(e). There will be an integration over a bit interval of the message signal $m(t)$, as also shown in Figure 7-8(e). Finally, decision-making follows in the manner described in the previous section. The result shows that the original transmitted data is now recovered at the receiver.

7.4 Pseudo-random Binary Sequence

This section may be too advanced for many computer science students. It may be ignored if that is the case.

The spreading signal $c(t)$ in Equation (7-3) is generated based on a pseudo-noise (PN) binary sequence $\{c_k\}$ with a noise-like waveform. The sequence c_k is usually generated by means of a feedback shift register which consists of m flip-flops (two-state memory elements) and a logic circuit connected together to form a feedback circuit, as shown in Figure 7-9.

The logic circuit is generally formed by modulo-2 adders whose inputs are from the flip-flops of the shift register. The flip-flops are regulated by a single timing clock. At each pulse of the clock, the state of each flip-flop is shifted to the next one following the signal follow indicated by the arrows. The logic circuit computes from the state of the shift register and feeds back the result as the input to the first flip-flop. The output is taken from the right-most flip-flop. An example of a feedback shift register is shown in Figure 7-10.

In this example there are $m = 3$ flip-flops. The input, denoted as s_0, applied to the first flip-flop, is equal to the modulo-2 sum of s_1 and s_3. If the feedback logic is implemented by modulo-2 adders, the PN sequence generator is called a ***linear feedback shift register***. It is evident that the generated sequence in this case is periodic with the maximum possible period of $2^m - 1 = 7$. If the period of the sequence generated by the linear feedback shift register attains its maximum possible period, the sequence is called ***maximum-length sequence***. It can be verified that the sequence generated by the linear feedback shift register shown in Figure 7-10 is a maximum-length sequence that has period of 7. Let us assume that the initial state of the shift register is 100 (reading the contents of the three flip-flops from the left to the right). First, computing the feedback result s_0 from the modulo-2 adder, we have $s_0 = s_1 \oplus s_3 = 1 \oplus 0 = 1$. Now performing the shift operation – that is, assigning $s_1 = s_0$, $s_2 = s_1$ and $s_3 = s_2$ simultaneously – we have $s_1 = 1$, $s_2 = 1$ and $s_3 = 0$. Therefore, the state following the initial 100 is 110. Repeating the process, we have a succession of states as follows:

$$100, 110, 111, 011, 101, 010, 001, 100, \dots.$$

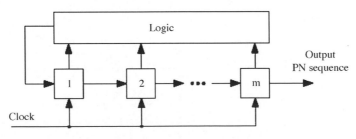

Figure 7-9   Feedback shift register

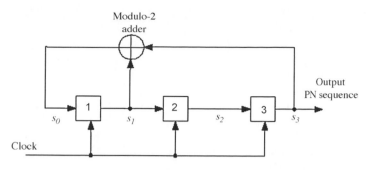

Figure 7-10   A maximum-length sequence generator for $m = 3$

The output sequence (the last position of each state of the shift register) is therefore $00111010\ldots$, which repeats itself with period $2^3 - 1 = 7$. Note that the choice of the initial state 100 is an arbitrary one. In addition, an all-zero initial state should be avoided;. otherwise the resulting sequence will be all zeros.

The length m can be extended arbitrarily so that the resulting sequence has a long period. For example, with $m = 30$, the resulting sequence will have a period of $2^{30} - 1 = 1073741823$. However, it should be noted that not all the linear feedback logics could generate a maximum-length sequence.

In Table 7-1 we give the set of linear feedback shift registers with length $m = 2, 3, \ldots, 8$ that generate a maximum-length sequence. We shall not give the proof of the correctness of this table. The reader should simply note that, using this table, pseudo-random sequences can be generated. The numbers given in the table indicate the flip-flops that should be connected in the feedback modulo-2 summation. For example, the numbers [5,4,2,1] given in the second entry of $m = 5$ indicate that the feedback connection should be taken from the fifth, fourth, second and first flip-flips as depicted in Figure 7-11. It can be verified that the generated sequence from Figure 7-11 is a maximum-length sequence.

Table 7-1   Maximum-length sequence of shift-register with lengths 2 through 8 feedback taps

m	Feedback taps
2	[2,1]
3	[3,1]
4	[4,1]
5	[5,4,3,2], [5,4,2,1], [5,2]
6	[6,5,3,2], [6,5,2,1], [6,1]
7	[7,5,4,3,2,1], [7,6,5,4,2,1], [7,6,5,2], [7,6,4,2], [7,3,2,1], [7,1]
8	[8,6,4,3,2,1], [8,7,6,1], [8,5,3,1], [8,6,5,2], [8,4,3,2]

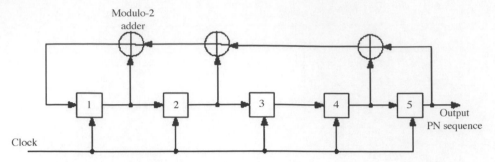

Figure 7-11   Maximum-length sequence generator with feedback connection [5,4,2,1]

7.5 Frequency-hopping Spread-spectrum

In the DSSS modulation discussed earlier, the use of a PN sequence to modulate a phase-shift-keyed signal achieves instantaneous spreading of the transmitted bandwidth. In this section we will introduce another spread-spectrum technique called *frequency-hopping spread spectrum*. Unlike with DSSS techniques, the frequency-hopping spread spectrum spreads sequentially rather than instantaneously. *That is, the carrier frequency of the transmitted signal hops randomly from one frequency to another so that the spectrum of the transmitted signal covers only one small channel at any instance, but covers a wide range in the long run*.

Since frequency hopping does not cover the entire spread spectrum instantaneously, we are led to consider the rate at which the hops occur. There are two basic types of frequency hopping:

- *Slow-frequency hopping*, in which the information symbol rate is larger than the hopping rate; several symbols are transmitted on one frequency hop.
- *Fast-frequency hopping*, in which the information symbol rate is smaller than the hopping rate; the carrier frequency will change several times during the transmission of one information symbol.

In this section, for simplicity, we will consider only the slow-frequency hopping technique.

Figure 7-12 shows the block diagram of a frequency-hop transmitter and receiver. The incoming data are fed into a digital modulator with a fixed carrier frequency f_x. The digital modulator may be one of the digital modulators introduced in Chapter 5. A common modulation technique for frequency-hopping systems is FSK. A frequency synthesizer is used to synthesize another carrier signal with a desired frequency f_s.. This frequency is controlled by k-bit segments of a PN sequence which enables the carrier frequency to hop over 2^k distinct values. The resulting modulated wave from the digital modulator and the output from the frequency synthesizer are then fed into a mixer that consists of a multiplier followed by a band-pass filter.

The filter is designed to pass the sum frequency component resulting from the multiplication process as the transmitted signal. For example, assume that the carrier

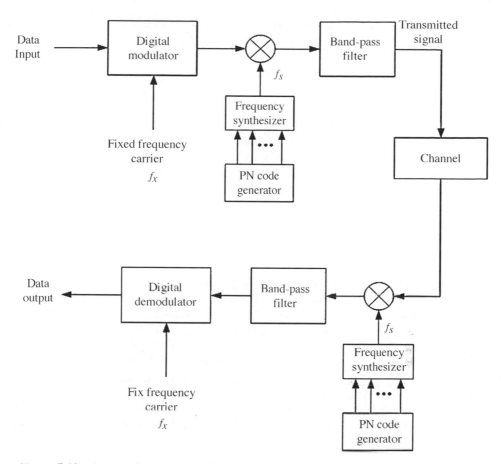

Figure 7-12   A transmitter and a receiver of the frequency hopping spread spectrum system

frequency generated from the synthesizer is f_s. Then the carrier frequency after the mixer is $f_x + f_s$. ***Remember that the multiplication of two cosine functions results in the addition of the frequencies of these two functions. This is why a multiplier is used in the mixer.***

The bandwidth of the transmitted signal, on a single hop, is the same as that resulting from the use of a conventional digital modulator. However, for a complete range of 2^k frequency hops, the transmitted signal occupies a much larger bandwidth than that of the information-bearing signal.

In the receiver in Figure 7-12, the frequency hopping is first removed by mixing (down-converting) the received signal by multiplying the received signal with the output from a local frequency synthesizer. The frequency synthesizer is synchronously controlled in the same manner as in the transmitter. The resulting signal is then applied to a band-pass filter to remove unwanted signal and noise. The filter is designed to pass the subtracted frequency component resulting from the multiplication process. For example, assume that the received signal has carrier frequency $f_r = f_x + f_s$. Now assume that the carrier frequency generated by the local frequency synthesizer is also f_s. After the multiplication, the signal becomes

$$\cos(2\pi(f_x + f_s)t))\cos(2\pi f_s t)$$
$$= \frac{1}{2}(\cos(f_x + 2f_s)t) + \cos(2\pi f_x t)).$$

The carrier frequency after the band-pass filter will be f_x, which is fixed as expected. From now on, the carrier frequency of the signal after the mixer in the receiver is fixed. Therefore, a conventional digital demodulator can now be employed to recover the transmitted data.

7.6 Application of Spread-spectrum Techniques to Multiple-access Systems

In Chapter 6 we introduced various multiple-access systems. The spread-spectrum techniques can be used in such systems.

Let us first consider the CDMA system. As we indicated in Section 6.3, this requires that the signals for code are orthogonal to each other. Now, suppose we use randomly generated sequences consisting of 1s and −1s. It is easy to see that the inner product of any pair of these sequences tends to give a rather small number. Let these two sequences be denoted as $S_1 = (s_{11}, s_{12}, \ldots, s_{1n})$ and $S_2 = (s_{21}, s_{22}, \ldots, s_{2n})$. Consider the product of s_{1i} and s_{2i}. The probability of the product being 1 is equal to the probability of it being −1. Thus when we sum these products, the 1s tend to cancel out with the −1s, and the inner product − although not necessarily 0 − is usually quite small. We may therefore say that they are orthogonal.

These techniques can all be used with other techniques in a multi-access environment. For instance, in the IEEE.802.11 standard for wireless local-area networks, the direct-sequence spread-spectrum technique is used together with CSMA/CA. Yet, all users use the same sequence.

The Bluetooth standard is an interesting one. There are a master and several slaves. At any time, the master only talks to one slave using one particular frequency. The frequencies are determined by the frequency-hopping mechanism. The situation can be considered to be the same as a time-sharing operating system where the computer talks to terminals in turn. In the following, we will introduce the IEEE 802.11 and Bluetooth standards.

7.6.1 Introduction to the IEEE 802.11 Standard – DSSS Technology

The IEEE 802.11 is a wireless local-area network (WLAN) standard that operates in the 2.4 GHz band which was adopted by IEEE as the first standard for WLANs. Originally, the IEEE 802.11 WLAN was designed to look and feel like a wired LAN so that it appears to be the same as the wired networks to users. It supports all of the protocols and all of the LAN management tools that operate on a wired network. In this section we will focus on the modulation scheme of the standard.

First we have to understand the characteristics of the WLAN channels. There are a number of differences between wired LANs and WLANs. For example, the data on a WLAN are broadcast for all to hear. Secure communication thus becomes an important issue in WLAN systems. The air link also constitutes a more complex channel than that of the wired

LAN. Everything in the environment is either a reflector or an attenuator of the signal. This can cause significant changes in the strength of a signal received by a WLAN receiver. At the frequency band used in the IEEE 802.11 WLAN, the wavelength is

$$\lambda = v/f_c = (3 \times 10^8)/(2.4 \times 10^9) = 0.125\,\text{m} = 12.5\,\text{cm}.$$

Therefore, small changes in position can cause large changes in the received signal strength. In the wireless environment, this phenomenon is due to the signal traveling many paths of different lengths to arrive at the receiver. Each arriving signal is of a slightly different phase from all of the others. All the signals sometimes add up in phase and sometimes out of phase. The overall signal strength is hence sometimes large and sometimes small. Objects moving in the environment, such as people, doors, cars and other objects, can also affect the strength of the signal at a receiver by changing the attenuation or reflection of the many individual signals.

The IEEE 802.11 standard was originally designed to support 1 Mbps and 2 Mbps data rates and was later extended to support 5.5 Mbps and 11 Mbps data rates in IEEE 802.11b.

One of the IEEE 802.11 standards uses a direct-sequence spread-spectrum (DSSS) radio in the 2.4 GHz band. IEEE 802.11 was then extended to other high-rate standards (802.11a and 802.11g) using the orthogonal frequency division multiplexing (OFDM) technique. IEEE 802.11a/g supports data rates from 6 Mbps to 54 Mbps for the OFDM case. The bandwidth is around 20 MHz. In the following, we will describe the modulation scheme for the 1 Mbps and 2 Mbps schemes using DSSS.

Figure 7-13 shows the 1 Mbps DSSS modulator. The system is based on differential BPSK (DBPSK) modulation. At any time, our input is a_k which is either 1 or 0. Our first task is to transform a_k into b_k, which is 1 or -1. In the following, we shall call b_k an output bit. The input binary bit stream consisting of the a_k is first differentially encoded using the delay element and the XOR gate. (The term 'differentially' will be made clear later.) The sequences a_k and b_k are clocked at the information rate 1 Mbps. Then the output sequence can be expressed as

$$b_k = b_{k-1} \oplus a_k. \tag{7-14}$$

The above equation indicates that if the input $a_k = 1$ then $b_k = \overline{b_{k-1}}$, where $\overline{b_{k-1}}$ denotes the complement of b_{k-1}. If the input $a_k = 0$, then $b_k = b_{k-1}$. For the receiver to recognize a_k, it just needs to detect the phase difference between adjacent symbols. That is, if the phase of b_{k-1} is different from that of b_k, $a_k = 1$; otherwise, $a_k = 0$. This is why this kind of encoding is called differential encoding. The sequence b_k is then transferred to levels of $+1$ and -1 by

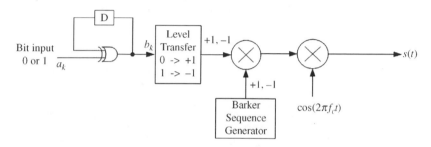

Figure 7-13   1 Mbps DSSS modulator

assigning $+1$ for binary symbol '0' and -1 for binary symbol '1', resulting in a bipolar sequence.

The reader may wonder why this kind of differential encoding is used. Suppose otherwise. A channel may for some reason reverse the polarity of the signals. In this case, the receiver may consequently decode the input signal incorrectly. With this kind of differential coding, the receiver always checks whether the phase of b_{k-1} is different from that of b_k. If they are different, $a_k = 1$; otherwise, $a_k = 0$.

The bipolar sequence is then multiplied by a length-11 sequence (called the Barker sequence) given by

$$+1, -1, +1, +1, -1, +1, +1, +1, -1, -1, -1.$$

The output of the multiplier results in a signal with a chip rate that is 11 times higher than the information rate. The result in the frequency domain is a signal that is spread over a wider bandwidth. Finally the signal is BPSK-modulated by multiplying the sequence by the carrier $\cos(2\pi f_c t)$, where f_c is around 2.4 GHz. Figure 7-13 shows a schematic diagram of the 1 Mbps DSSS modulator for the IEEE 802.11 standard.

Figure 7-14 illustrates a Barker-modulated signal where the upper signal is the output of the differential encoder and the lower signal is the Barker-modulated signal.

For the ideal case, the receiver given in Figure 7-5 can be used to receive the IEEE 802.11 signal, except that the block for the PN sequence generator is replaced by a Barker sequence generator. However, for a real wireless channel, the receiver design is in general more complex than that shown in Figure 7-5. Since the received signal in a wireless environment is a superposition of several arriving waveforms, each with different delay, the current received symbol may be interfered with by several previous symbols. This phenomenon is called ***intersymbol interference*** (ISI). The receiver design for IEEE 802.11 involves the technology of 'equalization', which is beyond the scope of this book.

For 2 Mbps modulation, a differential QPSK (DQPSK) modulation scheme is employed. Figure 7-15 gives the block diagram of the modulator. The input bit stream is first grouped into 2-bit blocks denoted as $(d_{0,k}, d_{1,k})$ at time index k. The 2-bit blocks are then differentially encoded by two delay elements and a 2-bit adder which adds the current input bits to previous output bits, resulting in a stream of 2-bit blocks $(b_{0,k}, b_{1,k})$. The output of the differential encoder is then mapped to a two-dimensional vector $(a_{0,k}, a_{1,k})$ of $+1$ and -1 by Table 7-2. The $(a_{0,k}, a_{1,k})$ forms one of the points in the QPSK constellation. The output sequence $a_{0,k}$ is then multiplied by a Barker sequence forming the signal in the so-called I-channel. The sequence of $a_{1,k}$ is also multiplied by the same Barker sequence forming the signal in the so-called Q-channel. The I and Q signals are then multiplied by

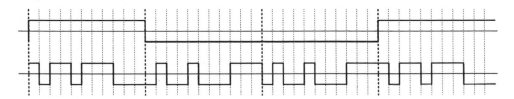

Figure 7-14 The sequence before and after using the Barker sequence multiplier

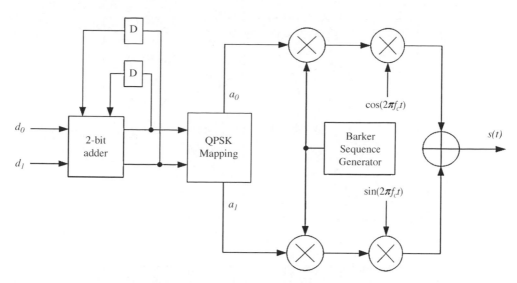

Figure 7-15 The 2 Mbps DSSS modulator

the carriers $\cos(2\pi f_c t)$ and $\sin(2\pi f_c t)$, respectively. Finally, they are added together to form the final transmitted signal.

The IEEE 802.11 adopted CSMA, a multiple-access technology which was introduced in Section 6.4. Under the CSMA mechanism, if an IEEE 802.11 transmitter wants to transmit a packet, it has to listen to the channel. If the channel is clean, then the transmitter can transmit the packet. If the channel is occupied, the transmitter holds the packet and tries to transmit the packet later.

7.6.2 Introduction to the Bluetooth Standard – Frequency Hopping Technology

The IEEE 802.11 standard is mainly used for a personal computer which provides wireless communication over the range of hundreds of meters. Bluetooth is a standard developed for short-range communications, up to 10 m, over the 2.4 GHz band. Bluetooth allows any sort of electronic equipment, from computers and cellphone to keyboard and headphones, to make its own connection without wires.

Bluetooth operates in the 2.4 Ghz band. In the US, a bandwidth of 83.5 MHz is available. In this band, 79 RF channels spaced 1 MHz apart are defined. The channel is used by a

Table 7-2 The QPSK mapping table

$(b_{0,k}, b_{1,k})$	$(a_{0,k}, a_{1,k})$
(0,0)	(1,1)
(0,1)	(1,–1)
(1,0)	(–1,1)
(1,1)	(–1,–1)

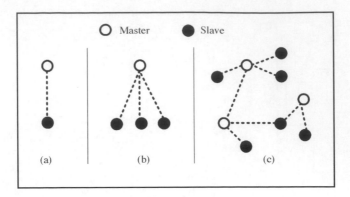

Figure 7-16   (a) Piconet with a single slave operation. (b) Piconet with multi-slave operation. (c) Scatternet operation

pseudo-random hopping sequence hopping through the 79 channels. That is, the carrier frequency of each channel is 2.4 GHz plus or minus, say, k megahertz. The frequency changes at a rate of 1600 hops per second. The bit rate is 723 Kbps. Two or more Bluetooth devices communicating with each other form a piconet as illustrated in Figures 7-16(a) and 7-16(b). **There is one master and one or more slave(s) in each piconet.** Multiple piconets with overlapping coverage areas form a scatternet as illustrated in Figure 7-16(c). Each piconet can have only a single master. However, slaves can participate in different piconets on a time-division multiplex basis. The hopping sequence is unique for the piconet and is determined by the master. The channel is divided into time slots where each slot corresponds to an RF hop frequency.

   The Bluetooth transceiver uses a time-division duplex (TDD) scheme, meaning that it alternately transmits and receives in a synchronous manner. The time slot for each transmission or receiving is 625 microseconds. For a TDD scheme, the master and slave alternatively transmit. For example, the master may start its transmission in even-numbered time slots only and the slaves may start their transmissions in odd-numbered time slots only. Figure 7-17 illustrates a TDD scheme used in the Bluetooth master and slave. The function

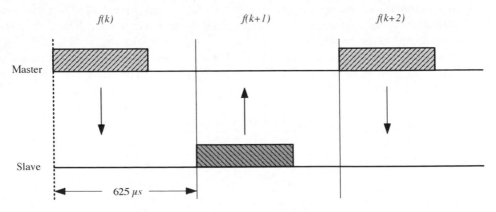

Figure 7-17   TDD for a master and a slave

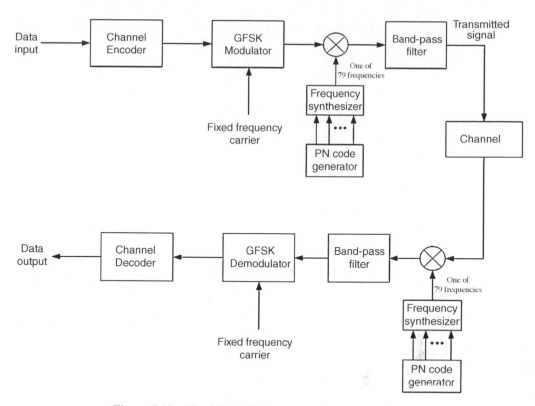

Figure 7-18   Simplified Bluetooth transmitter and receiver

$f(k)$ denotes the carrier frequency for transmitting/receiving of the kth slot. The master first sends a signal to its slave by using frequency $f(k)$; at the next time slot, the slave sends a signal to its master by using frequency $f(k+1)$; and so on.

Bluetooth adopts frequency hopping (FH) technology with **GFSK**. The term 'GFSK' stands for Gaussian frequency-shift keying which is a technology similar to FSK but with required transmission bandwidth less than that of the FSK. The data rate for the GFSK modulation is 1 Mbps. Figure 7-18 illustrates a simplified Bluetooth transmitter and receiver. The information data are first encoded by a channel encoder. The channel code used in the Bluetooth is either a (3,1) repetition code or a (15,10) Hamming code. The channel codes will be introduced in Chapter 8. The coded sequence is then applied to a GFSK modulator with a fixed frequency carrier. The modulated signal is then converted to one of the 79 carrier frequencies by a mixer. A frequency synthesizer is used to synthesize another carrier signal with desired frequency. This frequency is controlled by k-bit segments of a PN sequence which enables the carrier frequency to hop over 79 distinct values. The resulting modulated wave from the digital modulator and the output from the frequency synthesizer are then fed into a mixer that consists of a multiplier followed by a band-pass filter. The filter is designed to pass the sum frequency component resulting from the multiplication process as the transmitted signal.

In the receiver depicted in Figure 7-18, the frequency hopping is first removed by mixing (down-converting) the received signal by multiplying the received signal with the output from a local frequency synthesizer. The frequency synthesizer is synchronously controlled in the same manner as in the transmitter. The resulting signal is then applied to a band-pass filter to remove unwanted signal and noise. A GFSK demodulator is then employed to recover the coded bits. The GFSK demodulator is similar to the FSK demodulator. The demodulated signal is then fed into the channel decoder which can correct one error for the (15,10) code and also one error for the (3,1) repetition code. The concept of error-correcting codes will be introduced in Chapter 8.

Further Reading

- On the history of the development of spread-spectrum communications, see [S82].
- The basic concepts of spread-spectrum modulation can be found in [S77] and [P82].
- For further details on the pseudo-random sequence, see [G67].

Exercises

7.1   Consider the signal with the following frequency spectrum.
Suppose we multiply the signal with $\cos(2\pi f_c t)$. Will this enlarge the bandwidth of the signal?

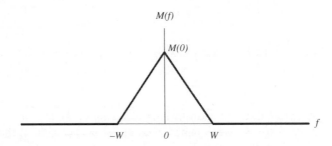

7.2   Use MATLAB to see how the change of the width of a pulse changes the bandwidth of the pulse.

7.3   For the DSSS method, suppose the PN signal $c(t)$ contains many narrow pulses. Will this induce a wider spectrum?

7.4   Suppose that you want to have a digital radio station which broadcasts music. Is the DSSS technology useful for this? Explain.

7.5   Many people like to use wireless earphones to listen to their mobile phone. Many of these systems use Bluetooth technology? Why?

7.6   Many computers use wireless keyboards. If communication between the keyboard and the computer uses a fixed frequency and there are many computers close to one another, will there be a problem? What would you do to avoid the problem?

8

Source Coding and Channel Coding

A common characteristic of signals generated by physical sources is that they contain a significant amount of information that is redundant. The transmission of the redundant information is therefore wasteful for the communication resources. For efficient utilization of resources, the redundant information should be removed from the signal prior to transmission. The process is called *source coding*, and there are two schemes:

- Operation with no loss of information is called data compaction or lossless data compression, in which the original data can be reconstructed from its coded data without loss of information.
- Operation with loss of information is called data compression, of which the reconstructed data may be different from its original one.

The data compression scheme is often used in the processing of analogue signals (e.g. audio and video), when a certain degree of signal distortion is allowable. In this chapter, for source coding, we will focus on the lossless data compression scheme and will not discuss data compression.

For example, assume that the binary data contains a long list of consecutive 0s. The transmitter has to repeat transmitting the 0s many times without coding. The efficiency of such systems can be improved by substituting the consecutive 0s with some specific code words. For example, if the number of consecutive 0s is 100, then the transmitter may just transmit some information to inform the receiver that there are 100 zeros in the subsequent data stream. For the source encoder to be efficient, we need knowledge of the statistics of the source. If some source symbols are know to be more probable than others, then we may use this feature in the generation of a source code. The idea is very simple – assigning short code words to frequent source symbols and long code words to rare source symbols. Such a coding scheme is called *variable-length code*.

Morse code is an example of a variable-length code. Letters of the alphabet are encoded into streams of marks and spaces, denoted as dots and dashes, respectively. We may consider these dots and dashes as 0s and 1s, respectively, as for digital communication systems. Statistical analysis of literature in the English language shows that the letter 'e' occurs more frequently then the letter 'q'. Therefore, the Morse code encodes 'e' into a single dot, the shortest code word in the scheme, and it encodes 'q' into $- - -$, the longest code word in the scheme.

Communications Engineering.   R.C.T. Lee, Mao-Ching Chiu and Jung-Shan Lin
© 2007 John Wiley & Sons (Asia) Pte Ltd

In a practical communications system, there is another important problem, namely that digital information may not be correctly received by the receiver. This phenomenon may be caused by noise or other impairments of the communication channel. It is important for a digital communication system to provide a certain level of reliability. For an unreliable channel, the only practical option available to increase the reliability is to use channel coding, termed 'error control coding'. The channel coding adds redundancy according to a prescribed rule. For example, we may encode information bit 0 as the coded bits '000' and 1 as '111', of which two redundant bits are added which are exactly the replica of the information bit.

Figure 8-1 shows a typical communications system that employs both source coding and channel coding. The discrete source generates information in the form of binary symbols. The redundant information of the source is first removed by the source encoder. The channel encoder in the transmitter accepts message bits from the source encoder and adds redundancy according to the channel coding rule. The coded bits are then mapped to the channel symbols by a modulator and transmitted over the channel. At the receiving end, the received signal may be corrupted by the channel noise which is generally the primary source of decoding error. The received signal is then demodulated by a demodulator and then passed to the channel decoder. The channel decoder exploits the redundancy to correct certain information errors; that is, it decides which bits were actually transmitted. The goal of the channel encoder and decoder is to minimize the effect of channel noise. That is, the number of errors between the channel encoder input and the channel decoder output is minimized. The information from the channel decoder is then fed into the source decoder, in which the original discrete source is recovered.

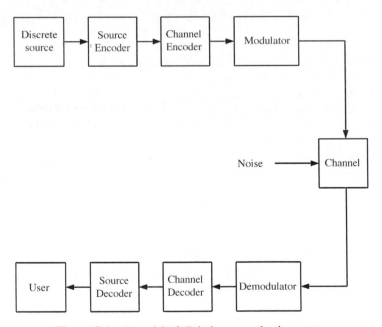

Figure 8-1 A model of digital communication system

There are numerous source and channel coding schemes. In this chapter, we focus on the concepts of coding and how the coding schemes work by examples rather than the coding theory. We will first introduce some source coding techniques and the most famous Huffman coding is used as an example. Then, we will introduce the channel coding techniques. For that, the Hamming code and convolutional code are used as examples.

8.1 Average Codeword Length of Source Coding

Consider a binary discrete source of alphabet $A = \{0, 1\}$ with statistics $\{p_0, p_1\}$. That is the probability of occurrence of a 0 is p_0 and that of a 1 is p_1. We assume that the symbols emitted by the source during successive signal intervals are statistically independent.

The source information bits are first grouped into blocks of length N, resulting in all 2^N possible binary words denoted as $A^{(N)}$. For example, if $N = 2$, then we have all the possible binary words of $A^{(2)} = \{00, 01, 10, 11\}$ with statistics $\{p_0^2, p_0 p_1, p_1 p_0, p_1^2\}$. We may denote $s_0 = (00)$, $s_1 = (01)$, $s_2 = (10)$ and $s_3 = (11)$. Now if $p_0 \neq p_1$, the probabilities of occurrence of elements of $A^{(2)}$ are different. For example, if $p_0 = 3/4$ and $p_1 = 1/4$, then we have the statistics $\{9/16, 3/16, 3/16, 1/16\}$. For variable-length source coding, we may design a set of code words of different lengths to represent the elements of $A^{(2)}$. Three possible source codes are shown in Table 8-1.

The encoding process simply assigns a unique code word to each element of $A^{(N)}$; that is, each element of $A^{(N)}$ is converted into a string of 0s and 1s. We assume that the source has an alphabet with $K = 2^N$ different symbols, and the kth symbol s_k occurs with probability p_k. Let the binary code word assigned to symbol s_k by the encoder have length l_k, measured in bits. For instance, in Table 8-1, the symbols are 00, 01, 10 and 11, with probabilities 9/16, 3/16, 3/16 and 1/16, respectivey. The length of each symbol is 2. The average code word length, $\bar{L}$, of the source encoder is defined as

$$\bar{L} = \sum_{k=0}^{2^N - 1} p_k l_k. \tag{8-1}$$

The parameter $\bar{L}$ represents the average number of bits per source symbol used in the source encoding process. We may also define the average length per information bit as

$$\bar{L}_b = \frac{1}{N} \sum_{k=0}^{2^N - 1} p_k l_k = \frac{\bar{L}}{N}. \tag{8-2}$$

Table 8-1 Illustration of three source codes

Source symbol	Probability of occurrence	Code I	Code II	Code III
$s_0 = (00)$	9/16	0	1	0
$s_1 = (01)$	3/16	00	10	10
$s_2 = (10)$	3/16	1	100	110
$s_3 = (11)$	1/16	11	1000	111

For example, from Table 8-1, if there is no source coding the digits from the source are directly transmitted without any process. In this case, the average code word length is

$$\bar{L} = 2 \times (9/16) + 2 \times (3/16) + 2 \times (3/16) + 2 \times (1/16) = 2.$$

The average length per information bit can be calculated as

$$\bar{L_b} = \frac{\bar{L}}{N} = \frac{2}{2} = 1.$$

Now assume that source coding is employed. For example, we assume that code III in Table 8-1 is used. Code III simply encodes the code word by the following mapping:

$$s_0 = (00) \rightarrow (0)$$
$$s_1 = (01) \rightarrow (10)$$
$$s_2 = (10) \rightarrow (110)$$
$$s_3 = (11) \rightarrow (111).$$

That is, it will output '0' if it encounters the source digits '00' and output '10' if it encounters the source digits '01', and so on. For example, if the original source bits are 001011010000, the encoder takes every two bits as a group and maps to the corresponding code word, resulting in the sequence 01101111000.

The average code word length of code III can now be computed as

$$\bar{L} = 1 \times (9/16) + 2 \times (3/16) + 3 \times (3/16) + 3 \times (1/16) = 27/16 = 1.6875.$$

The result shows that if code III is used, the average code word length can be reduced from 2 to 1.68; that is, we may use fewer coded bits to represent the original information bits. The average length per information bit for code III can be calculated as

$$\bar{L_b} = \frac{\bar{L}}{N} = 0.8438.$$

This result tells us the same thing, that we may on average use 0.8438 coded bits to represent one information bit.

8.2 Prefix Codes

For a source code to be of practical use, the code has to be uniquely decodable. We are specifically interested in a special class of codes satisfying a restriction known as the *prefix condition*. Any sequence made up of the initial part of the code word is called a prefix of the code word. For example, (0), (01), (011) and (0110) are all prefix of the code word (0110). A prefix code is defined as a code in which no code word is the prefix of any other code word.

Table 8-1 illustrates three source codes for encoding of $A^{(2)}$. Code I is not a prefix code, since the code word (0) for s_0 is a prefix of the code word (00) for s_1. Similarly, we may show

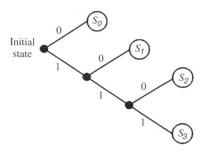

Figure 8-2   Decision tree for Code III of Table 8-1

that code II is not a prefix code, but code III is a prefix code with the important property that it is always uniquely decodable.

But the converse is not always necessarily true. For example, code II in Table 8-1 does not satisfy the prefix condition, yet it is uniquely decodable since the bit 1 indicates the beginning of each code word in the code.

To show that a prefix code is always uniquely decodable, we may take code III in Table 8-1 as an example. Assume that the original source bits are 001011010000. The encoder takes every two bits as a group and maps to the corresponding codeword, resulting in the sequence 01101111000. To decode the sequence of code words generated from the prefix code, the source decoder simply starts at the beginning of the sequence and decodes one code word at a time based on a decision tree. Figure 8-2 depicts the decision tree corresponding to code III in Table 8-1. The tree contains an initial state and four terminal states corresponding to source symbols s_0, s_1, s_2 and s_3. The decoder always starts at the initial state. The received bit moves the state up to the terminal state s_0 if it is 0, or else moves down to a second decision point. The decoder will repeat the process until the decoding state reaches the terminal state. Once the decoding state reaches the terminal state, the decoder will output the corresponding symbol of the terminate state. Then the decoding state is reset to its initial state and the whole decoding process is performed again until the end of the coded string.

8.3 Huffman Coding

We will describe an important class of prefix codes known as Huffman codes. The basic idea of Huffman coding is to assign the symbol of large probability a short code word and assign the symbol of small probability a long code word. The constructed codes are uniquely decodable since they are prefix codes. The essence of the algorithm is to construct a decision tree step-by-step. The construction of the Huffman code proceeds as follows:

1. The two source symbols of lowest probability are assigned 0 and 1.
2. These two least-probable source symbols are then combined into a new source symbol whose probability is equal to the sum of the probabilities of the original two symbols. The number of source symbols is decreased by one due to this combination.

The procedures of step 1 and step 2 are repeated until the number of source symbols is reduced to one.

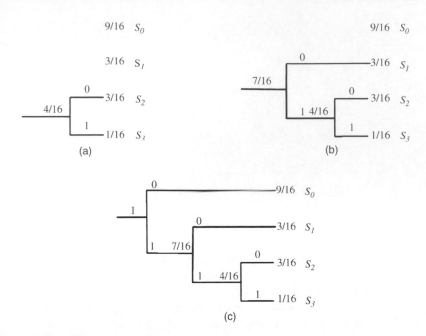

Figure 8-3 The construction of a Huffman code for 4 symbols, (a) the first step, (b) the second step, (c) the last step

For example, the construction of four-symbol Huffman code with source probabilities 9/16, 3/16, 3/16 and 1/16 is illustrated in Figure 8-3. The first step is shown in 8-3(a), of which the two least probable symbols with probabilities 3/16 and 1/16 are combined into one new symbol with probability $3/16 + 1/16 = 4/16$. The two least probable symbols are assigned by 0 and 1, respectively. The problem is now reduced to constructing a three-symbol Huffman code with source probabilities 9/16, 3/16 and 4/16. The two least probable symbols among the three symbols are with probabilities 3/16 and 4/16. Figure 8-3(b) shows the combination of symbols with probabilities 3/16 and 4/16 and the bit assignment, resulting in a new symbol with probability 7/16.

Now there are only two symbols left with probabilities 9/16 and 7/16. The final step is to combine these last two symbols and assign 0 and 1, respectively, to them. Now, we have constructed a tree. The code word for each source symbol can be read out from the left-most assigned symbol to the right following the tree. For example, for symbol s_3, the code word should be 111 read from left to right. For symbol s_2, the code word should be 110, and so on. The constructed code is exactly the same code as code III in Table 8-1. By using Equation (8-1), the average code word length is $\bar{L} = 1.6875$. The average length per information bit can be calculated as $\overline{L_b} = \bar{L}/2 = 0.8438$. This result shows that the efficiency of representing source bits can be improved significantly by Huffman coding.

To demonstrate how the source coding improves the efficiency of the digital communications system, we assume that there is a system with 1 Mbps transmission rate; that is, the channel encoder shown in Table 8-1 accepts 1 megabits per second from the source encoder. Now if no source encoder is employed, the transmission data rate is 1 Mbps. If the source has the statistics as shown above and Huffman coding is used, the transmission rate can be improved to be $(1/0.8438) = 1.1851$ Mbps.

8.4 Channel Coding

As we indicated before, unlike source coding, channel coding adds redundancy to increase the reliability of communication. Thus channel coding is also called *error-correcting code*. There are many different such codes proposed in the literature. Historically, these codes have been classified into *block codes* and *convolutional codes*. The distinguishing feature between them is the presence or absence of memory in the encoding. Convolutional codes have memory embedded in the encoders while block codes do not.

A block code accepts information in successive k-bit blocks. For each block, it adds $n - k$ redundant bits that are algebraically related to the k message bits. Therefore, it produces an overall encoded block of n bits, where $n > k$. The n-bit block is called a code word, and n is called the block length of the code. For simplicity, we will call such a block code (n, k) block code.

The error-correcting codes are particularly useful when the channel contains a high level of noise. Such noise causes very unreliable transmission. We say that a transmission bit is in error if the transmitter transmits 0 while the receiver receives 1, or if the transmitter transmits 1 while the receiver receives 0. Now if the probability of the bit error is large, the transmission of information is not easy without any channel coding.

A simple example of channel codes is repetition code. Each code word of the repetition code represents one information bit; that is, $k = 1$. The encoder simply repeats the information bit $n - k = n - 1$ times. For example, let $n = 3$. If we want to transmit a 1, then we may add two additional 1s as the redundant bits, resulting in the code word $(1,1,1)$; if we want to transmit a 0, then we may add two additional 0s as the redundant bits, resulting in the code word $(0,0,0)$. A repetition code of length n can correct at most $t = (n - 1)/2$ errors. In this case, $n = 3$, so we may correct one error within three coded bits. For example, assume that (111) was transmitted and the received word is (011). There is one error in the first position. Now, in the receiver, the decoder will find a code word, among (000) and (111), which is closest to the received word (011). The code word (000) has two positions that are different from the received word (011) and the code word (111) has just one position that is different from (011). Therefore, in this case, the decoder would decide that the transmitted code word is (111), and conclude that the decoded information bit is a 1. It is easy to verify that the decoder can correct one error within the three coded bits.

8.5 Error-correcting Capability and Hamming Distance

We have demonstrated a simple idea of error-correcting code by using repetition codes. Now, the problem is how to measure the error-correcting capability of a block code. The key to this question is the Hamming distance of a block code. Let $c_1 = (c_{1,1}, c_{1,2}, \cdots, c_{1,n})$ and $c_2 = (c_{2,1}, c_{2,2}, \cdots, c_{2,n})$ be two n-tuple binary vectors. The Hamming distance between c_1 and c_2, denoted as $H(c_1, c_2)$, is defined as the number of digits of c_1 and c_2 that are different. For example, the Hamming distance between (010) and (100) is 2, since the first and second positions are different.

Let C be a set of codewords. For example, for the repetition code of length 3, $C = \{(000),$ $(111)\}$. Note that a code may contain more then two code words. The minimum distance of C, denoted as d_{min}, is defined as the minimum Hamming distance between distinct codewords of C; that is:

$$d_{min} = \min_{\substack{c_1 \in C, c_2 \in C \\ c_1 \neq c_2}} H(c_1, c_2). \tag{8-3}$$

Example 8-1
Consider the following code (call it A):

$$0 \to 000$$
$$1 \to 111.$$

The d_{min} of this code is 3. Now consider another code (B), which is as follows:

$$00 \to 000$$
$$01 \to 011$$
$$10 \to 101$$
$$11 \to 110.$$

It can be seen that the d_{min} of this code is 2.

At the receiver, upon receiving a word x, the decoder will find a code word $D(x)$ among C that is closest in Hamming distance to x. Explicitly, the decoder should find the code word

$$c = D(x) = \arg\min_{c' \in C} H(x, c'). \tag{8-4}$$

This is called a minimum-distance decoder.

Example 8-2
Consider code A in the above example. Suppose the receiver receives 010, it will decode it as a 0. If it receives 110, it will decode it as a 1. Suppose now that code B is used and the receiver receives 001. The decoder cannot uniquely decode the message: it can say only that the original message is either 00, 01, or 10.

For an (n, k) block code, there are 2^k code words of length n, denoted as $c_1, c_2, \cdots, c_{2^k}$ in C. Define the set $B_i = \{x : D(x) = c_i\}$. It is obvious that, based on minimum-distance decoding, all the possible 2^n n-tuple combinations can be partitioned into 2^k regions, $B_1, B_2, \cdots B_{2^k}$, so that $B_i \cap B_j = \phi$, for $i \neq j$. The decoder will output code word c_i if and only if $x \in B_i$.

Intuitively, the larger the minimum distance of the code, the better the error-correcting capability of the code will be. Example 8-2 already informally shows this point. The error-correcting capability of a block code is defined as the maximum number of errors that can be corrected by the decoder. For a code of minimum distance d_{min}, the error-correcting capability is

$$t = (d_{min} - 1)/2. \tag{8-5}$$

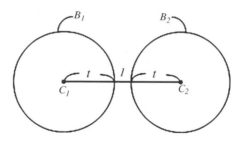

Figure 8-4   Minimum Hamming distance and error correcting capability

This can be explained by Figure 8-4, in which two minimum-distance code words, c_1 and c_2 of C, are depicted. The circles around the code words represent the regions B_1 and B_2, respectively. If the received word x is within B_1, which means that c_1 is the code word closest to x. In this case, the decoder will output the decoded codeword c_1. If the received word x is within B_2, then the decoder will output the decoded code word c_2. Since the Hamming distance between c_1 and c_2 is $d_{\min}$, the maximum number of errors that can be corrected within B_1 or B_2 should be $(d_{\min} - 1)/2$; otherwise B_1 and B_2 will overlap. For example, for code A in Example 8-1, which is a repitition code of length 3, the error-correcting capability of this code is 1, since $d_{\min} = 3$ in this case.

8.6 Hamming Codes

We will use a famous class of error-correcting codes, called Hamming codes, as an example. The length n and the number of information bits k of the Hamming code have the forms $n = 2^m - 1$ and $k = 2^m - 1 - m$ for some positive integer $m \geq 3$. The Hamming codes have the capability of correcting one error within n coded bits.

For example, let $m = 3$, so we have a (7,4) Hamming code. There are therefore 16 code words in the (7,4) Hamming code. Let (i_1, i_2, i_3, i_4) be the information bits. The code word of Hamming code has the form $(i_1, i_2, i_3, i_4, p_1, p_2, p_3)$, where p_1, p_2 and p_3 are the redundant bits that are added by the encoder. The redundant bits are also called *parity check bits*. The values of p_1, p_2 and p_3 are determined based on the encoding rule of the (7,4) Hamming code. The values of p_1, p_2 and p_3 must be added so that the following vectors contain an even number of 1s:

$$(i_1, i_3, i_4, p_1) \tag{8-6}$$

$$(i_1, i_2, i_3, p_2) \tag{8-7}$$

$$(i_2, i_3, i_4, p_3). \tag{8-8}$$

For example, if $(i_1, i_2, i_3, i_4) = (1, 0, 0, 1)$, we have $p_1 = 0$ to satisfy that the vector (i_1, i_3, i_4, p_1) contains an even number of 1s. We may call p_1 as the parity check bit for i_1, i_3 and i_4. Similarly, we have $p_2 = 1$ and $p_3 = 1$. Therefore, the encoded code word should be (1,0,0,1,0,1,1).

We will use Figure 8-5 to illustrate the encoding process. There are three circles representing the vectors of Equations (8-6), (8-7) and (8-8). For example, the top-most

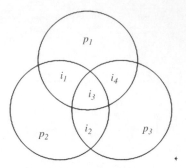

Figure 8-5   The diagram for Hamming code encoder

circle encloses the four components of (i_1, i_3, i_4, p_1). The values of p_1, p_2 and p_3 must be added so that each circle contains even number of 1s.

For another example, if $(i_1, i_2, i_3, i_4) = (1, 0, 0, 1)$, we may first fill the information bits in the corresponding positions as illustrated in Figure 8-5, resulting in the diagram in Figure 8-6. Now, we may decide the values of p_1, p_2 and p_3 based on the rule that every circle must contain an even number of 1s. The result is again $p_1 = 0$, $p_2 = 1$ and $p_3 = 1$.

All possible code words of the (7,4) Hamming code are given in Table 8-2 based on the encoding rule illustrated above. It can be verified that the minimum distance of this code is 3 by comparing all pairs of code words. Therefore, the (7,4) Hamming code can correct one error.

We will now demonstrate a decoding scheme and show how to correct one error upon receiving a word. Assume that the transmitted code word is $c = (1, 0, 0, 1, 0, 1, 1)$ and there is an error occurring in the second position, resulting in the received word $x = (1, 1, 0, 1, 0, 1, 1)$. The problem of decoding is how to locate the position of the error by simple parity check rules expressed in vectors of Equations (8-6), (8-7) and (8-8). Figure 8-5 may be used to help to

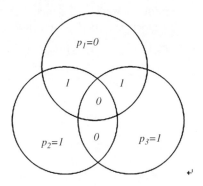

Figure 8-6   The encoding of $(i_1, i_2, i_3, i_4) = (1, 0, 0, 1)$

Table 8-2   Information bits and code words of (7,4) Hamming code

Information Bits	Codeword	Information Bits	Codeword
(0,0,0,0)	(0,0,0,0,0,0,0)	(1,0,0,0)	(1,0,0,0,1,1,0)
(0,0,0,1)	(0,0,0,1,1,0,1)	(1,0,0,1)	(1,0,0,1,0,1,1)
(0,0,1,0)	(0,0,1,0,1,1,1)	(1,0,1,0)	(1,0,1,0,0,0,1)
(0,0,1,1)	(0,0,1,1,0,1,0)	(1,0,1,1)	(1,0,1,1,1,0,0)
(0,1,0,0)	(0,1,0,0,0,1,1)	(1,1,0,0)	(1,1,0,0,1,0,1)
(0,1,0,1)	(0,1,0,1,1,1,0)	(1,1,0,1)	(1,1,0,1,0,0,0)
(0,1,1,0)	(0,1,1,0,1,0,0)	(1,1,1,0)	(1,1,1,0,0,1,0)
(0,1,1,1)	(0,1,1,1,0,0,1)	(1,1,1,1)	(1,1,1,1,1,1,1)

identify the parity check rules. We may first fill the received word $x = (1,1,0,1,0,1,1)$ according to the corresponding positions shown in Figure 8-5, resulting in the diagram shown in Figure 8-7. If there is no error, all circles should contain an even number of 1s. However, if one bit of the received word is in error, the number of 1s will not be even in some circles. Now we count the number of 1s among each circle of Figure 8-7. The topmost circle which corresponds to vector (8-6) contains an even number of 1s. We conclude in this stage that the bits among (i_1, i_3, i_4, p_1) are correct. The lower-left circle which corresponds to vector (8-7) contains an odd number of 1s. We conclude in this state that one bit among (i_1, i_2, i_3, p_2) is in error. The lower-right circle which corresponds to vector (8-8) also contains an odd number of 1s. We conclude in this state that one bit among (i_2, i_3, i_4, p_3) is in error.

In summary, we have:

- There is no error in (i_1, i_3, i_4, p_1).
- There is one error in (i_1, i_2, i_3, p_2).
- There is one error in (i_2, i_3, i_4, p_3).

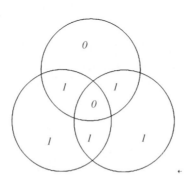

Figure 8-7   The decoding upon receiving the word $(1,0,0,1,0,1,1)$

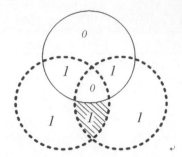

Figure 8-8   The identification of the error location

Remember that we have assumed that there is only one bit error in the received word. The problem now is how to identify the location of the error based on the above induction. In Figure 8-8, the circles that contain error are enclosed by dashed lines. We may find that the bits of the intersection between two circles in question are i_2 and i_3. However, from the above deduction (point 1), i_3 is correct. We may finally conclude that i_2 should be in error and correct this error by changing $i_2 = 1$ to $i_2 = 0$.

Table 8-3 shows all possible parities and the corresponding error positions. The decoder can simply count the number of 1s among vectors (8-6), (8-7) and (8-8) upon receiving a word and correct the error based on Table 8-3.

Equations (8-6) to (8-8) are not the only equations for the (7,4) Hamming code. Consider the following equation:

$$(i_2, i_3, i_4, p_1) \tag{8-9}$$

$$(i_1, i_3, i_4, p_2) \tag{8-10}$$

$$(i_1, i_2, i_4, p_3). \tag{8-11}$$

These three equations can be illustrated as in Figure 8-9 is quite a similar manner to Figure 8-6.

In general, every p_i should be associated with a 3-tuple consisting of i_1, i_2, i_3, and i_4 and 3-tuples corresponding to p_1, p_2 and p_3 must be distinct.

Table 8-3   Parities and error locations of (7,4) Hamming code

p_1	p_2	p_3	Error Position
Even	Even	Even	No error
Even	Even	Odd	7
Even	Odd	Even	6
Even	Odd	Odd	2
Odd	Even	Even	5
Odd	Even	Odd	4
Odd	Odd	Even	1
Odd	Odd	Odd	3

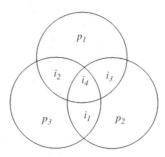

Figure 8.9   Another diagram for Hamming code encoder

8.7 Convolutional Codes

As noted in Section 8.4, convolutional codes differ from block codes in that the encoder contains memory. At any given time unit, the encoder accepts k information bits and outputs n coded bits. The n coded bits at any given time unit depend not only on the k inputs at that time unit but also on m previous input blocks. We will denote such a convoltional code as an (n, k, m) convoltional code.

A simple example of convolutional code is given in Figure 8-10. The input information sequence is one bit at any given time and is denoted by u_i at time index i. There are two output coded bits at any given time. The coded bits are denoted by $v_i^{(1)}$ and $v_i^{(2)}$. Two memory elements are used in the encoder for storing the previous information bits u_{i-1} and u_{i-2}. The coded bits $v_i^{(1)}$ and $v_i^{(2)}$ are functions of u_i, u_{i-1} and u_{i-2}, and are given by

$$v_i^{(1)} = u_i \oplus u_{i-2}$$
$$v_i^{(2)} = u_i \oplus u_{i-1} \oplus u_{i-2},$$

(8-12)

where the notation $\oplus$ denotes logical operator exclusive or (XOR). Since the encoder is a sequential circuit, it constitutes a finite state machine (FSM). Hence its behavior can be described by a state diagram.

Define the encoder state at time i as (u_{i-1}, u_{i-2}), which consists of the two message bits stored in the shift register. There are $2^m = 2^2 = 4$ possible states. At any time instant, the encoder must be in one of these states. The encoder undergoes a state transition when a message bit is shifted into the register. The input u_i will cause state transition from

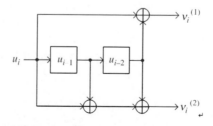

Figure 8-10   A (2,1,2) convolutional encoder

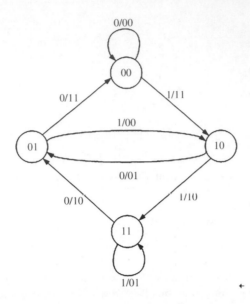

Figure 8-11 The state transition diagram of a (2,1,2) convolutional code

(u_{i-1}, u_{i-2}) to (u_i, u_{i-1}). The outputs $v_i^{(1)}$ and $v_i^{(2)}$ are obviously a function of current state (u_{i-1}, u_{i-2}) and the input u_i. Each state is represented by a vertex (or point) on a plane. The transition from one state to another is represented by a directed line.

Figure 8-11 illustrates the FSM of the encoder. Each branch is labeled by an input/output notation $u_i/v_i^{(1)}v_i^{(2)}$. For example, the label 1/01 means that the input information is $u_i = 1$ and the output is $v_i^{(1)}v_i^{(2)} = 01$.

The state transition diagram can also be illustrated by a code tree, as shown in Figure 8-12 for an input string of three bits.

In this tree, each bit above an edge is the input bit and the two bits below the edge are the output bits. Note that the output bits do not correspond to the input bit exactly, which is a special characteristic of convolutional code.

In practical applications, the length of the information bits is finite. Denote the length of the information bits by L. Let $(u_1, u_2, \ldots, u_L)$ be the information bits. The initial state of the encoder should be fixed to $(u_{i-1}, u_{i-2}) = (0, 0)$. For a better performance, the final state of the encoder should be also zero. Therefore, at the end of each information sequence, m zeros should be added to shift out the register so that the final state can be terminated to the zero state. In this example, since $m = 2$, two zeros are required to add at the end of the information sequence, resulting in the zero-padded information sequence $(u_1, u_2, \ldots, u_L, 0, 0)$. The zero-padded sequence is then input to the convolutional encoder. The output of the convolutional encoder can be expressed as

$$(v_1^{(1)}, v_1^{(2)}), (v_2^{(1)}, v_2^{(2)}), \ldots, (v_L^{(1)}, v_L^{(2)}), \underbrace{(v_{L+1}^{(1)}, v_{L+1}^{(2)}), \ldots, (v_{L+m}^{(1)}, v_{L+m}^{(2)})}_{\text{additional redundancy for zero padding}}.$$

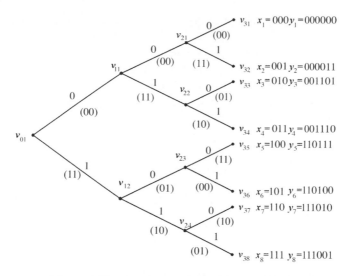

Figure 8-12 A code tree for a convolutional code

We may express the sequence by the following vector:

$$v = (v_1^{(1)}, v_1^{(2)}, v_2^{(1)}, v_2^{(2)}, \ldots, v_{L+m}^{(1)}, v_{L+m}^{(2)}).$$

Since, for general applications, L is very large compared with m, the overhead of the redundancy added at the end of the encoded sequence is not significant.

We now give a simple example with $L = 5$. Let $(u_1, u_2, \ldots, u_5) = (1, 0, 1, 1, 1)$. The zero-padded sequence is $(u_1, u_2, \ldots, u_5, 0, 0) = (1, 0, 1, 1, 1, 0, 0)$. From equation (8-12) or from Figure 8-11 with initial state $(0, 0)$, we obtain the coded sequence as $v = (1, 1, 0, 1, 0, 0, 1, 0, 0, 1, 1, 0, 1, 1)$.

Let us consider the case where we are sending two bits, namely 00, 01, 10 and 11. Since we stipulate that we have to return the state back to the initial state, we have to add 00 to every input string. That is, the input strings are now 0000, 0100, 1000 and 1100. The input strings and the corresponding code words are as follows:

$$0000 \rightarrow 00000000$$
$$0100 \rightarrow 00110111$$
$$1000 \rightarrow 11011100$$
$$1100 \rightarrow 11101011.$$

Imagine that we receive 00110111, we can easily find we have received a code word and there is no error. Suppose we receive 00111111, we can determine that the code word closest to the received word is 00110111 and we can therefore conclude that the sent bit string is 0100.

Every code word of a convolutional code can be represented as a directed path inside the state transition diagram and any directed path of length L constitutes a code word. The initial state of the path should be the zero state and the final state should also be the zero state. The state transition diagram is very useful for determining the performance of the convolutional code.

The performance of a convolutional code is characterized by the Hamming distance of the convolutional code. The most important distance measure for convolutional codes is the minimum free distance denoted by d_{free}. If the bit length L is fixed, we may view the convoltional code as a block code with kL input bits and $n(L+m)$ output bits. Let C_L be the set of all possible codewords of information length L. The free distance for C_L is defined as

$$d_{free} = \min_{v_1 \neq v_2 \in C_L} H(v_1, v_2), \tag{8-13}$$

where $H(v_1, v_2)$ is the hamming distance between v_1 and v_2. Denote the Hamming weight of a vector v by $W(v)$ which represents the number of nonzero elements of v. Then we can show that $H(v_1, v_2) = W(v_1 \oplus v_2)$. For example let $v_1 = (1, 0, 1, 1)$ and $v_2 = (0, 1, 1, 0)$. It is obvious $H(v_1, v_2) = 3$. Now let $v = v_1 \oplus v_2 = (1, 1, 0, 1)$; we also have $W(v_1 \oplus v_2) = W(v) = 3$.

Convolutional codes have the property that, if v_1 and v_2 belong to C_L, the vector $v = v_1 \oplus v_2$ also belongs to C_L, since convolutional code is linear in the sense of XOR. Therefore the free distance can be simplified to

$$d_{free} = \min_{\substack{v \in C_L \\ v \neq 0}} W(v). \tag{8-14}$$

The above equation indicates that the free distance of a convolutional code is the minimum weight of nonzero code words. The minimum weight of nonzero code words can be searched from the state transition diagram given in Figure 8-13. This figure is exactly the same as Figure 8-11, except that the labeling for input/output pair is replaced by its corresponding weight of coded bits. From the state transition diagram, we can find the minimum-weight nonzero code word is the path from (00) state to (10), (01) and return to the (00) state. The

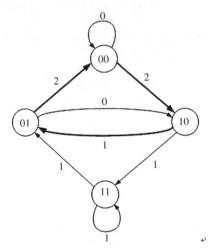

Figure 8-13   A state transition diagram with weight labeling

weight of the corresponding code word is $2 + 1 + 2 = 5$. Therefore, the free distance of this convoltional code is 5.

The Decoding Process

Imagine that the receiver receives a bit sequence $y = 111110$. The job of the decoding process is to determine the sent bit sequence whose length must be 3, in this case. To illustrate the process, let us first display the entire coding process for an input string of three bits in Figure 8-14.

As shown, there are $2^3 = 8$ possible sequences; and for each sequence x_i, its corresponding coded sequence y_i is shown. The bit above each edge is the input bit and the two bits below it are the corresponding output bits. Note that the output is not a function of the input only as the convolutional code has memory.

To determine the original input sequence, we may exhaustively search all of these eight output sequences and consequently determine that the distance between $y_7 = 111010$ and $y = 111110$ is 1 which is the smallest. Thus we conclude that the input sequence is 110. It should be made clear that there is an error in the transmitted bit sequence.

For each received sequence, we may change the tree in Figure 8-14 to a difference tree associated with this sequence. A typical example is shown in Figure 8-15. On each edge of the difference tree in Figure 8-15, we associate it with a cost which is the difference between the received two bits and the encoded two bits. For example, consider the edge $v_{01}v_{11}$. The encoded bits are 00 while the received bits are 11. Therefore the cost for this edge $v_{01}v_{11}$ is 2. On the other hand, for the edge $v_{01}v_{12}$, the encoded bits are 11 and the received bits are also 11. Consequently, the cost of the edge $v_{01}v_{12}$ is 0.

Having constructed a difference tree, the decoding problem becomes that of finding a shortest path from the root v_{01} to a lcaf node of the tree. In our case, the shortest path is $v_{01}v_{12}v_{24}v_{37}$ whose total is $0 + 1 + 0 = 1$ which is the smallest, among the costs of all paths. Again, this shortest path corresponds to the input sequence 110 which is correct.

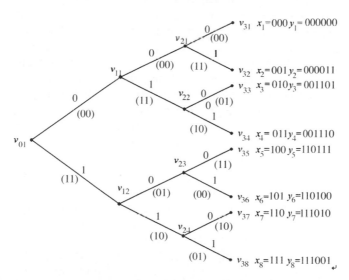

Figure 8-14   A tree illustrating the coding process involving a 3-bit input sequence

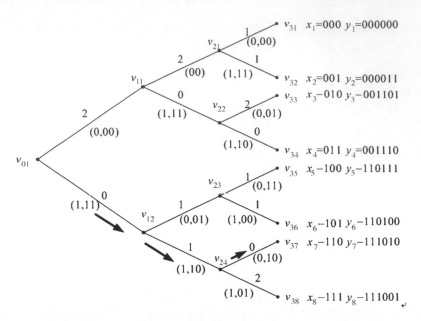

Figure 8-15 The difference tree for rceived sequence $y = 1\,1\,1\,1\,1\,0$

While this approach to decode a received sequence is straightforward and easy to implement, it faces a serious problem: the number of leaf nodes is exponential with respect to the length of the input sequence. For example, if the input sequence consists of 4 bits, the corresponding tree will have 16 leaf nodes. In general, if the length of the input sequence is k, there will be 2^k leave nodes and 2^k paths to search. Since our convolutional code does not put a limit on the length of the input sequence, this kind of approach is not practical.

Can we avoid an exponential exhaustive search? Yes, we can. Let us first exploit the notion of states. Consult Figure 8-11. In this figure, we can see that there are four states, namely (00), (10), (01) and (11). For convenience, let us label them as $a = (00), b = (11)$, $c = (01)$ and $d = (10)$. Then we redraw Figure 8-15 as Figure 8-16, emphasizing the fact that there are only 4 different states. Note that the graph in Figure 8-16 is not a tree any more, a characteristics which we shall use to decode a received message. We may conveniently call this a difference multi-stage graph.

Let us consider the node c_3. There are two paths terminating at this node, namely $a_0a_1b_2c_3$ and $a_0b_1d_2c_3$. The cost for $a_0a_1b_2c_3$ is $2 + 0 + 2 = 4$ and that for $a_0b_1d_2c_3$ is $0 + 1 + 0 = 1$. Thus we may ignore the path $a_0a_1b_2c_3$ because no matter what subsequences are from here, we will never use this path. Note that both paths $a_0a_1a_2b_3$ and $a_ob_1c_2b_3$ terminate at b_3. By the same reasoning, we may ignore the path $a_0a_1a_2b_3$ because the cost of this path is $2 + 2 + 1 = 5$ which is larger than the cost of path $a_ob_1c_2b_3$ which is $0 + 1 + 1 = 2$. Based on the above discussion, we may redraw Figure 8-16 as a simplified difference tree as in Figure 8-17; paths that can be ignored are now ignored.

Note that there are only four leaf nodes in the difference tree in Figure 8-17. We can easily see that from then on, using the same reasoning, we can always reduce the difference graph into a difference tree with only four leaf nodes. Figure 8-18 shows the case by extending the

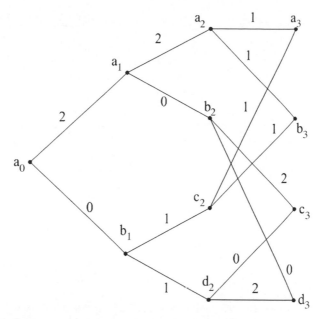

Figure 8-16 A multi-stage graph for the input sequence $y = 1\,1\,1\,1\,1\,0$

tree another stage. The reader can easily see that there are two paths terminating at, say b_4, and one of them can be ignored.

The above reasoning is the principle used in the Viterbi algorithm for convolutional code decoding. Actually, this is the dynamic programming strategy which is a commonly used

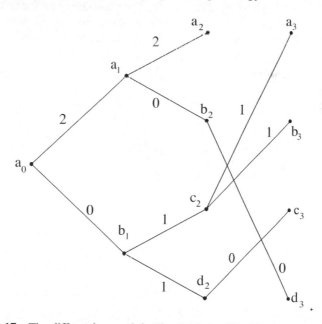

Figure 8-17 The differendce graph in Fig- 8-16 simplified by ignoring some paths

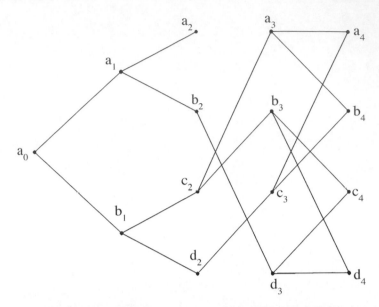

Figure 8-18   The tree in Fig- 8-17 extended another stage

strategy for algorithm design. Many algorithms are based on this approach. We shall not elaborate on dynamic programming in this book. The reader may vaguely understand the principle as follows: Suppose we want to find a shortest path from s to t. Suppose there are two nodes, say x and y, such that we may start from s, go to x and then finally to t. Similarly, we may also start from s, go to y and then to t. There are different paths from s to x and different paths from s to y. We do not know in the final solution which node should be used. Dynamic programming tells us no matter whether x is in the final solution or not; we should consider only the shortest path from s to x and ignore all of the paths from s to x. Similar arguments are valid for s and y.

Further Reading

- For more detail about source coding, see [B87].
- The Huffman code was invented by D. A. Huffman; see [H52].
- The Hamming code was invented by Hamming; see [H50].
- The convolutional code was invented by P. Elias; see [E55].
- For more detail about channel coding, see [LC83].

Exercises

8.1   Consider code III listed in Table 8-1. Decode the following two input sequences:

$$110010100111$$
$$01011110110.$$

8.2   Assume that there are seven messages with access frequencies 2, 5, 8, 15, 17, 19, 24. Use Huffman coding to code these seven messages.

8.3   Assume that the received messages are 0010111, 1101111 and 1010110. Determine the original sent messages under the assumption that Hamming code in Table 8-3 is used.

8.4   Assume that the state diagram of Figure 8-11 is used. Give the output sequences for the following input sequences: 1011, 1010 and 0101. Note that the input sequences have to be zero-padded.

8.5   Assume that the encoding process illustrated in Figure 8-13 is used and the received sequences are 111011 and 100100. Draw the difference trees corresponding to these sequences. Determine their original sequences.

Appendix

$$\cos(-\theta) = \cos(\theta) \tag{A-1}$$

$$\sin(-\theta) = -\sin(\theta) \tag{A-2}$$

$$\cos(\alpha + \beta) = \cos\alpha\cos\beta - \sin\alpha\sin\beta \tag{A-3}$$

$$\cos(\alpha - \beta) = \cos\alpha\cos\beta + \sin\alpha\sin\beta \tag{A-4}$$

$$\sin(\alpha + \beta) = \sin\alpha\cos\beta + \cos\alpha\sin\beta \tag{A-5}$$

$$\sin(\alpha - \beta) = \sin\alpha\cos\beta - \cos\alpha\sin\beta \tag{A-6}$$

$$a\cos\alpha - b\sin\alpha = \sqrt{a^2 + b^2}\cos(\alpha + \theta) \text{ where } \theta = \tan^{-1}\frac{b}{a} \tag{A-7}$$

$$a\cos\alpha + b\sin\alpha = \sqrt{a^2 + b^2}\cos(\alpha - \theta) \text{ where } \theta = \tan^{-1}\frac{b}{a} \tag{A-8}$$

$$\cos\alpha\cos\beta = \frac{1}{2}(\cos(\alpha + \beta) + \cos(\alpha - \beta)) \tag{A-9}$$

$$\sin\alpha\sin\beta = \frac{1}{2}(\cos(\alpha - \beta) - \cos(\alpha + \beta)) \tag{A-10}$$

$$\sin\alpha\cos\beta = \frac{1}{2}(\sin(\alpha + \beta) + \sin(\alpha - \beta)) \tag{A-11}$$

$$\cos\alpha\sin\beta = \frac{1}{2}(\sin(\alpha + \beta) - \sin(\alpha - \beta)) \tag{A-12}$$

$$\sin\alpha\cos\alpha = \frac{1}{2}\sin(2\alpha) \tag{A-13}$$

$$\cos^2\alpha - \sin^2\alpha = \cos 2\alpha \tag{A-14}$$

$$1 - 2\sin^2\alpha = \cos 2\alpha \tag{A-15}$$

$$2\cos^2\alpha - 1 = \cos 2\alpha \tag{A-16}$$

$$\sin^2\alpha = \frac{1}{2}(1 - \cos 2\alpha) \tag{A-17}$$

$$\cos^2\alpha = \frac{1}{2}(1 + \cos 2\alpha) \tag{A-18}$$

$$e^{j\theta} = \cos\theta + j\sin\theta \tag{A-19}$$

Communications Engineering. R.C.T. Lee, Mao-Ching Chiu and Jung-Shan Lin
© 2007 John Wiley & Sons (Asia) Pte Ltd

$$e^{-j\theta} = \cos\theta - j\sin\theta \tag{A-20}$$

$$e^{j\theta} + e^{-j\theta} = 2\cos\theta \tag{A-21}$$

$$\cos\theta = \frac{1}{2}(e^{j\theta} + e^{-j\theta}) \tag{A-22}$$

$$e^{j\theta} - e^{-j\theta} = 2j\sin\theta \tag{A-23}$$

$$\sin\theta = \frac{1}{2j}(e^{j\theta} - e^{-j\theta}) \tag{A-24}$$

Bibliography

[AR05]   H. Anton and C. Rorres, *Elementary Linear Algebra*, 9th ed., New York, John Wiley & Sons, 2005.

[AS65]   R. R. Anderson and J. Salz, 'Spectra of digital FM,' *Bell System Tech J.*, **44**, 1165–1189, 1965.

[AD62]   E. Arthurs and H. Dym, 'On the optimum detection of digital signals in the presence of white Gaussian noise: a geometric interpretation and a study of three basic data transmission systems,' *IRE Transactions on Communication Systems*, **CS-10**, 336–372, 1962.

[B00]   J. A. C. Bingham, *ADSL, VDSL, and Multicarrier Modulation*. New York, John Wiley & Sons, 2000.

[B65]   R. Bracewell, *The Fourier Transform and Its Applications*, 2nd ed., New York, McGraw-Hill, 1965.

[B87]   R. E. Blahut, *Principles and Practice of Information Theory*. Reading, MA, Addison-Wesley, 1987.

[BBC87]   S. Benedetto, E. Biglieri and V. Castellani, *Digital Transmission Theory*. Englewood Cliffs, NJ, Prentice-Hall, 1987.

[C85]   L. J. Cimmi Jr, 'Analysis and simulation of a digital mobile channel using orthogonal frequency division multiplexing', *IEEE Transactions on Communications*, **COM-33**, 665–675, 1985.

[C86]   A. B. Carlson, *Communication Systems*, 3rd ed., New York, McGraw-Hill, 1986.

[C93]   L. W. Couch, II, *Digital and Analog Communication Systems*, 4th ed., New York, Macmillan, 1993.

[C98]   G. E. Carlson, *Signal and Linear System Analysis*, New York, John Wiley & Sons, 1998.

[CF37]   J. R. Carson and T. C. Fry, 'Variable frequency electric circuit theory with application to the theory of frequency modulation,' *Bell System Tech J.*, **16**, 513–540, 1937.

[CL91]   E. F. Casas and C. Leung, 'OFDM for data communication over mobile radio FM channels,' *IEEE Transactions on Communications*, **COM-39**, 783–793, 1991.

[CL99]   L. J. Cimmi, Jr and Y. Li, 'Orthogonal frequency division multiplexing for wireless communications,' tutorial notes, *International Conference on Communications*, Vancouver, Canada, June 1999.

[E55]   P. Elias, 'Coding for noisy channels,' *IRE Convention Record*, pt 4, 37–46, March 1955.

[G67]   S. W. Golomb, *Shift Register Sequence*. San Francisco, Holden-Day, 1967.

[G93]   J. D. Gibson, *Principles of Digital and Analog Communications*, 2nd ed., New York, Macmillan, 1993.

[GHW92]   R. D. Gitlin, J. F. Hayes and S. B. Weinstein, *Data Communications Principles*. New York, Plenum, 1992.

[H00]   S. Haykin, *Communication Systems*, 4th ed., New York, John Wiley & Sons, 2000.

[H50]   R. W. Hamming, 'Error detection and error correcting codes,' *Bell System Tech. J.*, **29**, 147–160, 1950.

[H52]   D. A. Huffman, 'A method for the construction of minimum redundancy codes,' *Proceedings of the IRE*, **40**, 1098–1101, 1952.

[H59]   K. Henney (ed.), *Radio Engineering Handbook*. New York, McGraw-Hill, 1959.

[L95]   I. Lebow, *Information Highways and Byways*. Piscataway, NJ, IEEE Press, 1995.

[LC83]   S. Lin and D. J. Costello, Jr, *Error Control Coding: Fundamentals and Applications*, Englewood Cliffs, NJ, Prentice-Hall, 1983.

[LSW68]   R. W. Lucky, J. Salz and E. J. Weldon, Jr, *Principles of Data Communication*. New York, McGraw-Hill, 1968.

[NP00]   R. van Nee and R. Prasad, *OFDM Wireless Multimedia Communications*. Boston and London, Artech House, 2000.

Communications Engineering.   R.C.T. Lee, Mao-Ching Chiu and Jung-Shan Lin
© 2007 John Wiley & Sons (Asia) Pte Ltd

[OWY83]   A. V. Oppenheim, A. S. Willsky and I. T. Young, *Signals and Systems*. Englewood Cliffs, NJ, Prentice-Hall, 1983.

[P62]   A. Papoulis, *The Fourier Integral and Its Applications*. New York, McGraw-Hill, 1962.

[S77]   R. A. Scholtz, 'The spread spectrum concept,' *IEEE Transactions on Communications*, **COM-25**, 748–755, 1977.

[S79]   K. S. Shanmugam, *Digital and Analog Communication Systems*. New York, John Wiley & Sons, 1979.

[S82]   R. A. Scholtz, 'The origins of spread spectrum,' *IEEE Transactions on Communications*, **COM-30**, 822–854, 1982.

[S90]   F. G. Stremler, *Introduction to Communication Systems*, 3rd ed. Reading, MA, Addison-Wesley, 1990.

[SD92]   M. K. Simon and D. Divsalar, 'On the implementation and performance of single and double differential detection schemes,' *IEEE Transactions on Communications*, **COM-40**, 278–291, 1992.

[T95]   A. S. Tanenbaum, *Computer Networks*, 2nd ed. Englewood Cliffs, NJ, Prentice-Hall, 1995.

[V98]   S. Verdú, *Multiuser Detection*. Cambridge, Cambridge University Press, 1998.

[Z04]   K. S. Zigangirov, *Theory of Code Division Multiple Access Communication*. Wiley–IEEE Press, 2004.

[ZT02]   R. E. Ziemer and W. H. Tranter, *Principles of Communications: Systems, Modulation, and Noise*. Boston, Houghton Mifflin, 2002.

[ZW95]   W. Y. Zou and Y. Wu, 'COFDM: an overview,' *IEEE Transactions on Broadcasting*, **B-41**, 1–8, 1995.

Index

Communications Engineering. R.C.T. Lee, Mao-Ching Chiu and Jung-Shan Lin
© 2007 John Wiley & Sons (Asia) Pte Ltd